Aerospace Technologies and Applications for Dual Use

Aerospace Technologies and Applications for Dual Use

A New World of Defense and Commercial in 21st Century Security

General Pietro Finocchio

Prof. Ramjee Prasad

Prof. Marina Ruggieri

LONDON AND NEW YORK

Published 2008 by River Publishers

River Publishers

Alsbjergvej 10, 9260 Gistrup, Denmark

www.riverpublishers.com

Distributed exclusively by Routledge

4 Park Square, Milton Park, Abingdon, Oxon OX14 4RN

605 Third Avenue, New York, NY 10158

First published in paperback 2024

Aerospace Technologies and Applications for Dual Use A New World of Defense and Commercial in 21st Century Security / by General Pietro Finocchio, Prof. Ramjee Prasad, Prof. Marina Ruggieri.

Routledge is an imprint of the Taylor & Francis Group, an informa business

Publisher's Note

The publisher has gone to great lengths to ensure the quality of this reprint but points out that some imperfections in the original copies may be apparent.

While every effort is made to provide dependable information, the publisher, authors, and editors cannot be held responsible for any errors or omissions.

ISBN: 978-87-92329-04-2 (hbk)

ISBN: 978-87-7004-567-4 (pbk)

ISBN: 978-1-003-33713-3 (ebk)

DOI: 10.1201/9781003337133

To our capable and dedicated team behind this book

Table of Contents

Preface

इन्द्रियस्येन्द्रियस्यार्थे रागद्वेषौ व्यवस्थितौ ।
तयोर्न वशमागच्छेत्तौ ह्यस्य परिपन्थिनौ ॥

indriyasyendriyasyārthe rāga-dveṣau vyavasthitau
tayor na vaśam āgacchet tau hy asya paripanthinau

Attraction and aversion of the senses to their corresponding sense objects is unavoidable; one should not be controlled by them; since they are obstacles in one's path.

—The Bhagvad-Gita (3.34)

The aim of the Book is to present — for the very first time — an advanced, broad and multi-source view on Aerospace Technologies and Applications for Dual Use.

Book editors and chapter authors exchanged their ideas during an International Event, organised and sponsored by AFCEA (Armed Forces Communications and Electronics Association) Rome Chapter, and technically co-sponsored by IEEE AESS (Aerospace and Electronic Systems Society), IEEE Systems Council and IFIP (International Federation for Information Processing) that took place in Rome (Italy) on September 12–14, 2007.

Moving from the major outcomes of Sessions and Panels developed during the Event, the authors — who belong to civil and military institutions, industries and universities — have translated their thoughts and conclusions into the Chapters of this Book that, hence, represents a precious tool for shaping effectively the future of Dual Use.

Figure P.1 illustrates the coverage of this book. The book is divided into 4 parts: namely, Part 1 deals with Trends and Challenges in Aerospace Dual Use; Part 2 explores Dual Use Technologies; Part 3 discusses Dual Use Applications and finally Part 4 reports the Industry Outlook on Dual Use.

Since this book has been completed with the help of several contributions-from various types of organisations we have formatted the book so that the information is presented in a coherent manner. There are four chapters (5, 25, 28 and 31) where information has been provided in a presentation format.

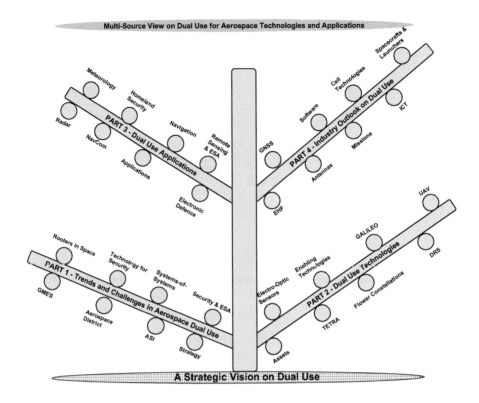

Figure P.1 Coverage of the Book

Curriculum Vitae of the Editors

Lieutenant General Pietro FINOCCHIO was born in Naples on 31 March 1945; He attended the Pozzuoli Aeronautical Academy from1963 to 1969 as a cadet and; He obtained a degree in Electronic Engineering from the Polytechnic of Naples, achieved with the maximum marks possible.

He was promoted to the rank of Major General of the Italian Air Force on 31st December 1996. Worked for the following organisation within the Italian Air Force:

- Technical Aeronautical Telecommunications Surveillance Office of Selenia, the Italian electronic manufacturing company;
- Italian Air Force Experimental Flight Test Centre (RSV), Pratica di Mare;
- Costarmaereo (MOD Procurement Agency for Aeronautical Weapon Systems);
- Office of the Aeronautical and Defence Attaché at the Italian Embassy in Washington;
- Telecomdife (MOD Procurement Agency for Air Defence, Air Traffic Control, Telecommunication and Informatics).

He carries the title of Avionics System Flight Test Engineer obtained in the Experimental Flight Test Centre in 1977, carries a civil pilot's licence and is an experienced parachutist. He has held the following positions in the course of his career:

- Host Country Inspector for the NATO programme for integrated air defence (NADGE);
- Technical Director of Experimental Flight Test Centre (RSV) of the IAF;
- Head of Section for the Procurement of Electronic Warfare Systems at Costarmaereo;
- Director of 3rd level — Tornado Aircraft maintenance Centre;
- Head of EFA and AMX Logistics — within Logistic branch of Italian Air Force General Staff;
- Head of 2° Department in Costarmaereo — responsible for the Procurement of aircraft, engines and avionics on behalf of the Italian Air, Navy and Ground Forces;

- Head of 1° Department in Costarmaereo — responsible for aircraft research and development projects;
- Co-Chairman of the EH101 Anglo-Italian Government Management Committee and member of the EH101 Policy Group, from 1990 to 1996;
- Head of the Italian delegation to the Joint Executive Committee and member of the Steering Committee of the NH90 Programme, from 1992 to 1996;
- Director of the Future Large Aircraft programme at Costarmaereo and member of the Policy Group, from 1994 to 1996;
- Director of the Technical Coordination Office of the General Directorate for Telecommunication, Air Defence and Traffic Control of Ministry of Defence, from 1996 to 1999
- Chief of National Armament Directorate R&T Department from 1999 to 2001.

Lieutenant General Chairman of Western European Armament Group Panel II(R&T) and R&T Committee of Western European Armament Organization from 20 June 1997 to June 2001.

He is General Manager of NAHEMA (NATO Helicopter Management Agency) from 1 August 2001 to 31 July 2004.

Since 27 September 2004 he is General Manager of TELEDIFE and from January 2005 he is the President of AFCEA (Armed Force Communications Electronics Association) ROME Chapter.

He was promoted to the rank of Lieutenant General on the 3 of February 2008. He speaks fluently English, French and Portuguese.

Lt. Gen. FINOCCHIO is married with Mrs Enrica and has a daughter and a son.

Ramjee Prasad was born in Babhnaur (Gaya), India, on July 1, 1946. He is now a Dutch citizen. He received his B.Sc. (eng.) from the Bihar Institute of Technology, Sindri, India, and his M.Sc. (eng.) and Ph. D. from Birla Institute of Technology (BIT), Ranchi, India, in 1968, 1970, and 1979, respectively.

He joined BIT as a senior research fellow in 1970 and became an associate professor in 1980. While he was with BIT, he supervised a number of research projects in the area of microwave and plasma engineering. From 1983 to 1988, he was with the University of Dares Salaam (UDSM), Tanzania, where he became a professor of telecommunications in the Department of Electrical Engineering in 1986.

At UDSM, he was responsible for the collaborative project Satellite Communications for Rural Zones with Eindhoven University of Technology, The Netherlands.

From February 1988 through May 1999, he was with the Telecommunications and Traffic Control Systems Group at DUT, where he was actively involved in the area of wireless personal and multimedia communications (WPMC). He was the founding head and program director of the Center for Wireless and Personal Communications (CWPC) of International Research Center for Telecommunications — Transmission and Radar (IRCTR).

Since June 1999, Dr. Prasad has been with Aalborg University, where currently he is Director of Center for Teleinfrastruktur (CTIF), and holds the chair of wireless information and multimedia communications. He is coordinator of European Commission Sixth Framework Integrated Project MAGNET (My personal Adaptive Global NET). Beyond MAGNET he was involved in the European ACTS project FRAMES (Future Radio Wideband Multiple Access Systems) as a DUT project leader. He is a project leader of several international, industrially funded projects. He has published over 500 technical papers, contributed to several books, and has authored, coauthored, and edited 25 books, the most recent being: *Ad Hoc Networking towards Seamless Communications, 4G Roadmap and Emerging Communication Technologies* and *Applied Satellite Navigation Using GPS, Galileo and Augmentation Systems*. His current research interests lie in Wireless networks, packet communications, multiple access protocols, advanced radio techniques, and multimedia communications. He has several patents to his credit.

Dr. Prasad has served as a member of the advisory and program committees of several IEEE international conferences. He has also presented keynote speeches, and delivered papers and tutorials on WPMC at various universities, technical institutions, and IEEE conferences. He was also a member of the European cooperation in the scientific and technical research (COST-231) project dealing with the evolution of land mobile radio (including personal) communications as an expert for The Netherlands, and he was a member of the COST-259 project. He was the founder and chairman of the IEEE Vehicular Technology/Communications Society Joint Chapter, Benelux Section, and is now the honorary chairman. In addition, Dr. Prasad is the founder of the IEEE Symposium on Communications and Vehicular Technology (SCVT) in the Benelux, and he was the symposium chairman of SCVT'93. Presently, he is the Chairman of IEEE Vehicular Technology Communications / Information Theory Society Joint Chapter, Denmark Section.

In addition, Dr. Prasad is the coordinating editor and editor-in-chief of the *Springer International Journal on Wireless Personal Communications* and a member of the editorial board of other international journals. He was the technical program chairman of the PIMRC'94 International Symposium held in The Hague, The Netherlands, from September 19–23, 1994 and also of the Third Communication Theory Mini-Conference in Conjunction with GLOBECOM'94, held in San Francisco, California, from November 27–30, 1994. He was the conference chairman of the fiftieth IEEE Vehicular Technology Conference and the steering committee chairman of the second International Symposium WPMC, both held in Amsterdam, The Netherlands, from September 19–23, 1999. He was the general chairman of WPMC'01 which was held in Aalborg, Denmark, from September 9–12, 2001, and of the first International Wireless Summit (IWS 2005) held also in Aalborg, Denmark on September 17–22, 2005.

Dr. Prasad was also the founding chairman of the European Center of Excellence in Telecommunications, known as HERMES, and now he is the Honorary Chair. He is a fellow of IEE, a fellow of IETE, a senior member of IEEE, a member of The Netherlands Electronics and Radio Society (NERG), and a member of IDA (Engineering Society in Denmark). Dr. Prasad is advisor to several multinational companies. He has received several international awards; the latest being the "Telenor Nordic 2005 Research Prize" (website: http://www.telenor.no/om/).

Marina Ruggieri graduated in Electronics Engineering in 1984 at the University of Roma. She was: with FACE-ITT and GTC-ITT (Roanoke, VA) in the High Frequency Division (1985–1986); Research and Teaching Assistant at the University of Roma Tor Vergata (RTV) (1986–1991); Associate Professor in Telecommunications at Univ. of L'Aquila (1991–1994) and at RTV (1994–2000). Since November 2000 she is Full Professor in Telecommunications at RTV (Faculty of Engineering), teaching DSP and Information and Coding.

Since 2003 she directs a Master in "Advanced Satellite Communications and Navigation Systems" at RTV.

Her research mainly concerns space communications and navigation systems (in particular satellites), integrated systems as well as mobile and multimedia networks.

Since 1999 she has been appointed in the Board of Governors of the IEEE AES Society and, since 2005, Director for AESS Operations in Italy&Western Europe.

Since January 2008 she has been elected Executive Vice President of the IEEE AESS.

In 2004–2006 she has been member of Technical-Scientific Committee of the Italian Space Agency (ASI). Since July 2007 she is Vice-President of the Technical-Scientific Committee of ASI.

Since December 2007 she belongs to the Italian *Superior Council of Telecommunications* as Expert.

Since December 2006 she il Vice President of the *AFCEA Rome Chapter.*

She is the Italian representative in the Technical Committee *Communications Systems* (TC6) of IFIP (International Federation for Information Processing).

She is Director of *CTIF_Italy*, the Italian branch of the *Center for Teleinfrastruktur* (CTIF) in Aalborg (Danimarca), opened on September 28, 2006 at RTV.

She has been P.I. of various national Programs funded by ASI and MIUR, an Internalization Program funded by MIUR, an ESA Ariadna Program and she has coordinated the RTV Unit in various European Projects funded by EU and GALILEO Joint Undertaking.

She is Editor of the IEEE Transactions on AES for "Space Systems", Chair of the IEEE AES Space Systems Panel, Assistant Editor of the IEEE Aerospace and Electronic Systems Magazine. Since 2002, she is co-chair of Track 2 "Space Missions, Systems, and Architecture" of the AES Conference. She participates in the organisation of many international events.

She was awarded the *1990 Piero Fanti International Prize* and she had a nomination for the *Harry M. Mimmo Award* in 1996 and the *Cristoforo Colombo Award* in 2002.

She is author of about 250 papers, on international journals/transactions and proceedings of international conferences, book chapters and books.

She is an IEEE Senior Member (S'84-M'85-SM'94), an AFCEA, IIN and AICA Member.

List of Acronyms

ABBREVIATION	MEANING
ARTEMIS	Advanced Relay and TEchnology MIssion Satellite
ASA	Austrian Space Agency
ASI	Agenzia Spaziale Italiana — Italian Space Agency
ATHENA-FIDUS	Access on THeaters for European allied forces NAtions — French Italian Dual Use Satellite
BELSPO	Belgian Science Policy Office
COSMO-SkyMed	COnstellation of Small satellites for the Mediterranean basin Observation
CNES	Centre National d'Etudes Spatiales (France)
COTS	Commercial Off The Shelf
DERA	UK Defence Evaluation Research Agency
DRSS	Data Relay Satellite System
EGNOS	European Geostationary Navigation Overlay Service
EDA	European Defence Agency
ESA	European Space Agency
GAGAN	GLONASS And Geo-Stationary Augmented Navigation
GEOSS	Global Earth Observation System of Systems
GLONASS	Global Orbiting Navigation Satellite System
GMES	Global Monitoring for Environment and Security
GPS	Global Positioning System
INTA	Instituto Nacional de Técnica Aeroespacial (Spain)
IRIS	IP Routing in Space
ISINS	Italian Space Integrated Network for Security
LAAS	Local Area Augmentation System
MUSIS	MUltinational Space-based Imaging System
NASA	US National Aeronautics and Space Administration
NCO	Network Centric Operations
NCW	Network Centric Warfare
NGNs	Next Generation Networks
ORFEO	Optical and Radar Federation for Earth Observation
PNT	Positioning, Navigation and Timing
PRISMA	Precursore IperSpettrale Missione Applicativa — Hyperspectral Mission

PRS	Public Regulated Service (GALILEO)
SAR	Synthetic Aperture Radar
SBAS	Satellite Based Augmentation System
SDR	Software Defined Radio
SICRAL	Sistema Italiano per Comunicazioni Riservate e Allarmi — Italian System for Restricted Communications and Alarms
SNSB	Swedish National Space Board
SYNTHESIS	Synergies Generated by the Interoperability of Space Infrastructures for Public Regulated Services
SPOT	Satellites Pour l'Observation de la Terre — Earth-Observing Satellites
TDRSS	Tracking and Data Relay Satellite System
UAV	Unmanned Aerial Vehicle
VRS	Virtual Reference Station
WAAS	Wide Area Augmentation System
WGS84	World Geodetic System 84

Acknowledgements

Editors wish to thank all the people (organisers, speakers, panellists) that participated to the deployment of the International Event, and to all the authors of the Book.

A special recognition to Mirko Antonini and Attilio Vagliani for their enthusiastic effort in the completion of the Book.

PART 1

Trends and Challenges in Aerospace Dual Use

Chapter 1

Introduction: a Strategic Vision on Dual Use

Pietro Finocchio[*], Ramjee Prasad[**], Marina Ruggieri[***]

[*] *TELEDIFE / AFCEA*
[**] *University of Aalborg (Denmark) / CTIF*
[***] *University Rome Tor Vergata / AESS / AFCEA*

In today's global world the fourth dimension, that is Space, plays a key role, for military use as well as for civilian applications, both institutional and commercial: for this reason, for its two faces — i.e. both military and civilian — space technology has a "dual" characterisation. So, when dealing with "dual use systems", today, this definition addresses all systems conceived, designed and built to the end of allowing joint civilian and military use, for both civilian and military tasks.

With respect to Earth Observation systems, Italy is a leading country in the development of "dual" applications, having conceived, designed and built COSMO-SkyMed (COnstellation of Small satellites for the Mediterranean basin Observation), the first system in the world allowing simultaneous monitoring and remote sensing applications for civilian and military users, as the result of a successful co-operation between Italian civilian institutions and Defence (Figure 1.1).

Considering the paramount national interest, and in the light of "dual use" applications, in May 2007 the Italian Ministry of University and Research and the Ministry of Defence signed a Memorandum of Understanding for their co-operation in the domain of space research and development, in order to consolidate and reinforce their mutual relationship of consultation, co-ordination and co-operation affecting the technology and scientific research activities in the areas of common interest, so as to share and maximise the benefits of programmes, professional resources and competences, both in the national and in the international context, with special reference to "dual use" programmes and to the Italian participation in the European Space Agency and to the initiatives already in progress across the European Union. As a matter of fact, it is within the mission of the Italian Universities and the Research Public Institutions to set as centres for new ideas, scientific development, innovation and nurturing of professional skill essential to the growth of Italy in social, economic and cultural domains. For this reason, Universities and Research Public Institutions are key elements to the "Lisbon strategy", that is making Europe, by 2010, "the continent with the economy — based on knowledge — more dynamic and competitive in the world, able to implement a sustainable economic growth with new and better jobs, and improved social cohesion". In this view, the co-operation between the Research and Defence sectors in the conception of advanced systems

Figure 1.1 Artist's impression of COSMO-SkyMed

for dual use is a major stronghold in the development of Italy: this was already settled by tradition, but it now stands even more clearly thanks to this Memorandum of Understanding.

The Italian approach reflects today's world trend of considering systems, that already exist and are inter-dependent, in the perspective of a new definition in terms of "systems of systems", now by re-combination and integration — and in a future by re-consideration and re-design — to make them benefit one another in a wider context, enabling them to operate to many ends, with national and global consequences. This trend materializes, at European and world level respectively, in the international co-operation programmes GMES (Global Monitoring for Environment and Security) and GEOSS (Global Earth Observation System of Systems), where the international community plans to exchange, also with institutional users (regions, provinces, cities, towns), data supplied by Earth Observation satellites and Ground systems, for environmental control and security purposes.

Years ago, Italy already signed bilateral agreements with France, the major European Country acting in space activities in the domains of Earth Observation and Communications, in particular with aerospace technologies and applications for dual use.

In accordance with the Turin Agreement, signed in 2001 by the Italian and French Prime Ministers — Amato and Jospin — COSMO-SkyMed is part of the ORFEO (Optical and Radar Federation for Earth Observation) programme, which will enable a joint Italian-French observation capability through dual satellite sensors, optical (Pléiades) and radar (COSMO-SkyMed), and exchange of complementary satellite imaging products.

The Italian tradition of radar observation satellites is solid, and known worldwide; nevertheless, as for dual use applications, and in particular to manage environmental emergencies and disasters, the optical sensor is essential, it is strongly advisable that, in the medium term, Italy promote the scientific and industrial research on this type of sensor, developing new approaches as hyper-spectral technology (measuring the spectrum of an image to improve the information content), which has already been consolidated by the Italian industries in civilian space exploration programmes managed by the Italian Space Agency (ASI), the European Space Agency (ESA) and the National Aeronautics and Space Administration (NASA), such as Cassini, Rosetta, Venus Express, and Dawn. The political agreement of 2001 was followed by the agreement between the Italian Space Agency (ASI) and the matching French Agency CNES (Centre National d'Etudes Spatiales), and between the Italian Ministry of Defence and ASI. In the medium and long run, the co-operation between Italy and France will be enlarged at European level through the MUSIS (MUltinational Space-based Imaging System) programme, with the participation of Belgium, Germany, Greece and Spain. MUSIS aim is to implement a European multinational federated system for observation, recognition and reconnaissance from space, through a system made up of space constellations — national and international — that are different, but with a common and fully interoperable ground segment.

The dual use concept and the international co-operation gain ground also in the field of communications. As for Italy, the military satellite SICRAL 1, launched in 2001, will be soon followed by SICRAL 1B, with a new operational and contractual model allowing for the dual use approach, so that it will be possible to share the use of the satellite between military and civilian users, with applications dedicated to Carabinieri, Coast Guard, and Civil Protection, as well as telemedicine. Once again in the frame of the Italian-French co-operation, and with the involvement of their respective Ministries of Defence and Space Agencies, the ATHENA-FIDUS (Access on THeaters for European allied forces NAtions — French Italian Dual Use Satellite) communications programme is ongoing. As a logic complement to a dual use telecommunications integrated system, it is obvious to think of a system of satellites for data relay (DRSS — Data Relay Satellite System) able to operate in co-operation with other telecommunication systems, already in place or planned for the future. Such a system would allow Ground Stations to receive data from observation satellites (COSMO-SkyMed and GMES satellites in particular) within timeframes much shorter than the current ones, and with increased frequency. In the past, Europe already invested in such an experimental satellite (ARTEMIS — Advanced Relay and TEchnology MIssion Satellite), designed and built mainly in Italy. A system like this is of strategic importance, as it is an enabler for many applications, such as data transmission from remote probes (e.g. missions to other planets), links with the International Space Station and with automatic transport systems, links with UAVs and USVs (aircraft without a pilot in the atmosphere and in space), and support to missions to the Moon and Mars, both manned and unmanned. The United States of America, who since the sixties have a broad space programme which includes systems of this kind, today have available in orbit as many as nine data relay satellites (TDRSS — Tracking and Data Relay Satellite System).

In the Universities and in the Research Public Institutions, the design and development of micro-satellites for small-scale missions are very active. The know-how and the technologies obtained across these activities could be transferred to dual use applications, through fast and cheap experimentation. In particular, it could be interesting to address:

- issues dealing with sensors for remote sensing (for applications of fire and pollution detection, study and prevention of climate change), putting more effort on the experimentation of the hyper-spectral optical sensor for aircraft (SIM.GA) — made by the Italian industry — on military and civilian scenarios, at least to build up the data base that will be essential to interpret the satellite data which, hopefully, will be soon made available with the launch of the PRISMA (Precursore IperSpettrale Missione Applicativa) satellite of the Italian Space Agency (ASI);
- issues of communications, for emergency backup applications of ground systems (electrical power plants, command and control centres, central government institutions, data base).

Furthermore, together with — and not instead of — the big satellite systems, micro-satellites enable to set-up early warning applications (at national and international level), thanks to short revisit times (i.e. passing frequently on the same spot) even if offering lower sensing performance when compared with bigger satellites.

As of the navigation systems, the simplicity, precision and cheap price has made the GPS system, initially built in the USA for military applications, a system that is by now of broad civilian use worldwide. Nevertheless, the fact that the GPS is under the control of a single Nation, the USA, led many States to pursue the implementation of their own navigation systems, autonomous and under their own control: the Soviet Union, and then Russia, proceeded along this way, and today China and India are determined to follow as well. On its own, Europe started the GALILEO project, recently funded entirely with public money.

About this subject, it is vital to stress that navigation, as well as communications, is an enabler for many other applications: actually, the positioning of observation satellites is made simple and precise by the availability of navigation systems (actually, only GPS as of today), whereas "time critical" applications often rely on acquiring a time signal distributed via a satellite system. It is not to incur an infrastructure risk that Europe needs a navigation system on its own so as to have — as a complement to GPS — direct control and management over information which is critical and enabling for many other applications, that without this information would be much more expensive not to say impossible. In particular, in what concerns the dual approach to GALILEO, there is a specific navigation service, designated as PRS — Public Regulated Service, about which Italy should rapidly encourage the application research, especially concerning security, for Armed Forces, Civil Protection, Police, Customs and government institutions.

In the domain of communications, one of the emerging technologies, and most promising for the future, is the Software Defined Radio (SDR): whereas the traditional communication and navigation systems are mostly made up with specific hardware components, in the SDR the main operating functions are implemented through software applications called "wave forms", which makes them extremely flexible as the main characteristics can be modified through software, without any hardware replacement. This allows the achievement of very high levels of interoperability across different users and, moreover, the use of this technology improves the robustness of communications, in terms of integrity and privacy of data, with dual use (military/civilian) to the benefit of defence forces, civil protection, police and homeland security. In particular, in satellite applications, SDR technology allows interoperability among constellations of different types (communication, navigation, observation), without the need of providing dedicated systems on a case-by-case basis.

In order to be in a position to implement what highlighted so far, it will be possible — and it is advisable — to use the services provided by the VEGA launcher, mainly made by Italian companies, that is now in the development phase.

Together with the initiatives necessary for the development of products and space platforms, it is also essential to support research and development of basic technologies, which are common to many areas, enable state-of-the-art development and, by their own nature, are essentially "multi-use".

In addition to photonics and, at a later stage, nanotechnologies, extreme attention should be placed to the research and development of basic technologies for generation, amplification and transmission of signals. Amplifiers with ever increasing bandwidth and efficiency are essential for radio and television transmissions, mobile phone communications, different types of radars for various applications, satellite communications, SDR, and so on.

The technologies with highest efficiency, bandwidth, purity and, mainly, energy efficiency, at the moment are based on Gallium Arsenide (GaAs) and, in a future, will rely on Gallium Nitride (GaN): with very good reason they are both considered a strategic asset, and for this reason they are subject to restrictions and possible embargoes. In view of this, it is essential to mature, at national Italian and at European level, autonomous competences and technology capabilities.

With the contribution of the funds of the military research, and in co-operation with the major European industrial companies, it is already ongoing the study and the preliminary development of the complete technological process (microelectronics, packaging and, most important, foundry) for solid state amplifiers with very high power, bandwidth, purity and, especially, energy efficiency, based on Gallium Nitride (GaN). The activities allowing the move from the research to the development step, and then to the industrialization and to the production phases, need to be carefully supported, as they are essential to provide and grant independence to Italy and Europe with a technology that is fundamental to many areas and is an enabler to state-of-the-art developments.

In conclusion, there are many possibilities for development in the domain of space and, as proven in history — one for all the US space programme for the conquest of the Moon — an investment in technology — and in space technology in particular — is a massive challenge for the industry of any Nation, and it hits all the industrial sectors, from aerospace to electronics, from metals to chemicals, with an economic return on investment that can be estimated in the order of ten times — or even more — the initial investment. There is great expectation for the aerospace sector in a dual use dimension, for its beneficial effects in the national economy and in the citizens' quality of life.

Chapter 2
The Security-related Perspective of the European Space Agency

Giuseppe Morsillo

European Space Agency (ESA)

2.1 Introduction

Europe is increasingly called to endorse responsibility on a global level while facing major challenges linked to the growing dependence on a globally interconnected infrastructure and the rapidly evolving international security environment.

The European Security Strategy (2003) highlights those global challenges, pointing in particular to the emergence of more diverse, less visible and less predictable threats to European security, i.e. terrorism, proliferation, regional conflict and organised crime. Calling on Europe to be more active, coherent and capable in addressing those threats means first and foremost to develop both its strategic planning capabilities based on reliable information and improved overall situational awareness and its capability to effectively conduct actions in support of peace and international security.

The EU Council paper on 'ESDP[1] and Space' (2004) recognises space as multiple-use technology that provides decisive added value in responding to those objectives, enabling the civilian and military user to see, listen, communicate, locate and synchronize information at global level and with uninterrupted availability. It equally underlines the need in Europe for the sharing and pooling of space-related resources and maximum use of multiple-use technologies for civilian and military operations.

With the adoption of the European Space Policy by Ministers of twenty-nine ESA and EU Member States on 22 May 2007, Europe has given itself a comprehensive space policy that for the first time explicitly recognises the security and defence dimension and calls on the different actors to increase synergies and dialogue between national, intergovernmental and communitarian stakeholders in this domain.

---

[1] Since its creation in 1999, the European Security and Defence Policy (ESDP) thereby relies on a mix of civilian and military tools, including space assets, for crisis management and conflict prevention.

2.2 ESA-EU Cooperation Adding a New Dimension
to European Space

The common will to put space at the service of society and European policies has been at the basis of the Framework Agreement between ESA and the European Community (EC) in force since 2004. Creating the 'Space Council', i.e. the concomitant meeting of the ESA and EU Council at Ministerial level, as well as a Joint ESA/EC Secretariat, the Framework Agreement will continues to be the basis of cooperation between the Agency and the European Union.

Prior to the adoption of the Resolution on the European Space Policy by the Space Council on 22 May 2007, European Ministers had already made important steps in clarifying the roles and responsibilities of each party: The EU is tasked to federate user demand and take the lead in applications and development of related services in support of EU policies while ESA takes the lead in space research and development activities in science, exploration, launchers, technology and the space component of application programmes.

In support of such applications, the Earth Observation programme on Global Environment for Space and Security (GMES) and navigation, positioning and timing satellite system GALILEO were established as European flagship programmes.

2.3 The Added Value of Space in Support of Europe's Strategic
Objectives and Policies

Convergence among Member States, ESA and EC has been achieved on a variety of Strategic Objectives, recognising the strategic importance of space for Europe:

- Developing and exploiting space applications serving Europe's public policy objectives and needs of European enterprises and citizens, including in the field of environment, development and global climate change
- Meeting Europe's security and defence needs
- Ensuring a strong and competitive space industry which fosters innovation, growth and the development and delivery of sustainable and cost-effective services
- Contribute to the knowledge-based society (science and exploration)
- Securing unrestricted access to new and critical technologies, systems and capabilities

The security dimension is present across all strategic objectives as it addresses also issues of human security with regard to climate change, supports a

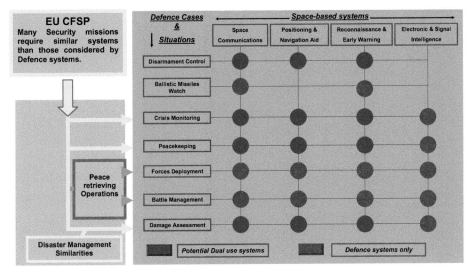

Figure 2.1 Potential dual use systems in the space domain

competitive European industry by cutting-edge innovation and secures the strategic non-dependence of Europe.

In addition, space assets are major enablers for capabilities needed for civilian and military operation as space-based sensors have the advantage of unrestricted access over potential areas of operation:

– **Space imagery** is crucial for the planning and conduct of operations assessing the situation on the ground, contributing to early warning, to conflict and proliferation prevention, and to treaties compliance verification.
– **Secure and reliable communications** are essential for exercising political control and strategic direction of any operation, especially in case of deficient or damaged infrastructure.
– **Space based positioning and time distribution systems** is indispensable to all military forces and civilian units.
– **At national level** space systems can also contribute to early-warning/ballistic missile defence and to signal intelligence (COMINT and ELINT).

2.4 Security and Programmatic Highlights in the European Space Policy

In response to the EU Council's call for the development of a global EU space policy, including agreed ESDP requirements, several programmatic activities highlighted

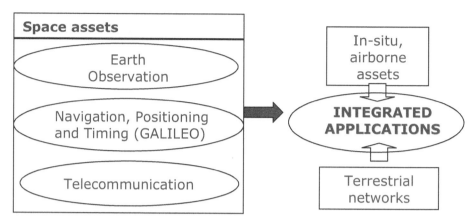

Figure 2.2 The concept of Integrated Applications

in the European Space Policy and in which ESA participates have a clear security dimension:

a) The support to European flagship application programmes

- **GMES** provides Europe with an independent capability for global monitoring and vital information on the global environment. It will also support Europe's needs for security (e.g. disaster monitoring, crisis management). The EU Council will identify the ESDP requirements relevant to GMES services.
- **GALILEO** is the first joint ESA/EU programme, providing independent capability for Positioning, Timing and Navigation services. It is of significant strategic importance both from a technology and an application point of view. It remains a civil programme under civilian control but may have military users.

b) The concept of Integrated Applications

The ESP highlights the need and potential for increasingly and seamlessly linking different space assets as well as in-situ and airborne assets in a system-of-systems approach. Such approach is of particular relevance to security-related applications in a network centred configuration.

c) Specific actions for the security and defence domain

Europe has the opportunity to improve coordination between defence and civilian space programmes, given their large technological commonality, while taking advantage of a more open and structured dialogue among national and European bodies. Specific attention has been given to the following aspects:

- **Set up a structured dialogue between the competent bodies** for optimizing synergies within the framework of existing attribution of competences.
- **Develop new security technologies and infrastructures** at European level for which strategic non-dependence is deemed crucial.
- **Protect European space-based capabilities against disruption** given that the security and economy of Europe and its citizens are increasingly dependent on them.

2.5 Developments in ESA Regarding Security-Relevant Activities

a) Overall activities

In line with its Convention[2] and based on actions identified in the European Space Policy, ESA has entered into a dialogue with major entities of the security and defence domain to explore opportunities of cooperation and synergy and build on a user-driven approach.

In parallel, the Agency has initiated the preliminary assessment of a number of security-relevant candidate priorities both in the technological and applications domain.

- **Earth Observation:** The Agency will consider the inclusion of security-related requirements as defined by the EU Council into the GMES space component.
- **Navigation:** The Agency is a major player in the development of the GALILEO space component and follow-up systems.
- **Technology:** The Agency will propose the setting-up of a full fledged technology programme called 'NewPro', including technological development activities in support of European non-dependence and security.
- **Space Situational Awareness (SSA):** The Agency will propose the setting-up of dedicated programmatic activities.
- **Telecom initiatives:** The Agency assesses the possibility of setting up dedicated programmatic activities in the field of Data Relay Satellites.

ESA is also interested in assessing the interests of user groups representing different communities from the civil and security sectors in other promising dual-use

---

[2]The Convention gives ESA the mandate 'to provide for and promote, for exclusively peaceful purposes, cooperation among European States in space research and technology and their space applications', the term 'peaceful' being understood as 'non-aggressive'.

technology areas. Candidate areas for such assessment might encompass the following topics:

- Hyperspectral sensors
- Integration of UAVs and satellites
- (Laser links, DRS)
- Small/nano satellites, configuration
- Flying
- Multi Functional Structures

- Impact shielding
- Cryptography
- Radiation hardening
- Protection filters for sensors
- HF power generation
- Nulling antenna
- Anti-spoofing anti-jamming
- ...

These developments should be supported by the establishment of an exchange mechanism with the users to identify their primary needs in the above mentioned domains as already experienced in the field of 'Space Situational Awareness'.

b) Space Situational Awareness — an example of a vocational dual-use initiative

The ESP underlines that space assets have become indispensable enablers for a wide spectrum of applications and that "the economy and security of Europe and its citizens are increasingly dependent on space-based capabilities which must be protected against disruption."

In their Resolution on the European Space Policy, Ministers highlighted in particular that "the space sector is a strategic asset contributing to the independence, security and prosperity of Europe and its role in the world. Europe cannot afford to lose out securing the benefits of space for its citizens and its policies so as to remain a major player, solve global problems and improve the quality of life". The Resolution further emphasised that "all of Europe's space activities contribute to the goals and fully respect the principles set out by the United Nations' 'Outer Space Treaty' and that Europe supports the on-going efforts of the United Nations' Committee on the Peaceful Uses of Outer Space (COPUOS) on the mitigation and prevention of space debris."

In line with those provisions and bearing in mind that the European independent access to space and its space based services depend on the capacity to safely operate the European space infrastructures, interest has been shown in ESA Member States for the establishment of a "Space Situational Awareness" as a common, multiple-use, European initiative. "Space Situational Awareness (SSA)" is preliminarily defined as the comprehensive knowledge of the population of space objects, of existing threats/risks, and of the space environment.

Such initiative would aim at identifying, preventing or at least limiting the risk of disruption of space-based services, building on measures linked to space surveillance, space weather, space-based detection, monitoring and identification of illicit activities in view of creating an overall space situational awareness. It would usefully complement the currently limited technological competence and capability available in some ESA Member States.

For the moment, Europe does almost entirely rely on the data provided by the United States Air Force for detecting, tracking and cataloguing of a certain percentage of space objects. Europe is not in a position to effectively acquire an overall Space Situational awareness and thus autonomously guarantee the continuous operation of critical European space infrastructure, needed to support its global policies and to fully live up to its legal obligations. Based on the interest expressed by national and European user communities, ESA has coordinated a *Preparatory action*, notably by establishing and coordinating a 'User Representatives'/Expert Group' with a view to develop and harmonize requirements, propose architectural solutions and identify technological gaps. The group involves specialists from public and private entities stemming from both the civilian and military sector.

The work of this group has been supported by three parallel industrial studies commissioned by ESA on the "Capability gaps concerning European Space Situational Awareness". Their findings with regard to technology gaps are to be further refined by ESA.

Among the results emerging from the *Preparatory action* and from the three studies in particular, is the fact that the most challenging issue to be addressed concerns the development of a suitable governance and data policy model, thereby answering the question of "who is allowed to accesses which data under which conditions".

Against that political and programmatic background and building on further consultation with its Member States, ESA intends to prepare a programme proposal regarding the development of a European SSA System to be submitted to the ESA Council on Ministerial Level in late 2008. The proposal should include preliminary architectural options meeting user requirements as well as technology projects as needed.

2.6 Outlook

An internal reflection on security-related activities and plans is currently on-going within ESA as a coordinated agency-wide effort. Its intention is to define — on the basis of a strategic analysis — objectives and areas of involvement for ESA in the security domain within short-term, medium-term and long-term horizons.

After an internal harmonisation process the outcome will be discussed with ESA Member States and is currently planned to be presented to the ESA Council by the end of 2007.

Further steps, such as the preparation of dedicated security-related programme proposals for ESA's Ministerial Conference will depend on the level of consensus reached within the ESA framework. In the medium to long term, ESA will have to further deepen and intensify the dialogue with other actors such as the EU, OCCAR or EDA as well as with relevant national entities in the frame of the implementation of the European Space Policy.

Chapter 3
Complex Systems for Dual Use Applications at the Italian Space Agency

Mario L. Cosmo

ASI — Italian Space Agency

3.1 Introduction

Today Security is a "Global Scale" Problem that requires to be monitored on a "Global Scale". As a consequence in 2003 EU has adopted the European Security Defence Policy to protect its citizen during crises and to prevent conflicts and in 2004 space has been considered an essential asset to meet the requirements. It is also a fact that the world economy, both at national and at international level, has become increasingly dependant on Space-borne Systems.

Dual-use Programs justify the big investments required today by Complex Space Programs by sharing costs, risks and successes between Government entities and eventual commercial partners. Space-based applications, as offered by an integrated use of Earth Observations, Telecoms and Navigation, have an inherent Dual Use Capability and hold great promise.

3.2 The National Scenario

When ground infrastructures, in emergency areas, are damaged or unavailable, space infrastructures may represent the unique alternative to guarantee rescue operations and the continuity of governmental action.

The Italian Space Agency (ASI) has been at the fore-front in the development of Dual-use Systems, including related products and their applications.

The National Space Plan includes Programs with Dual Capability as:

– Satellite Communications Technologies for mobile users;
– EO Satellite Systems (COSMO-SkyMed, SABRINA and the Hyperspectral Mission);
– Integrated EO-TLC-NAV Applications and Technologies for Institutional Users;
– GALILEO PRS Applications and Services.

Over the years ASI has been developing an infrastructure whose integration holds elements of great innovation in offering valuable services to Institutional Users. The model is also intended to be extended to European Partners as well.

COSMO-SkyMed can be considered the first building block toward the realization of this vision. In addition, ASI is a major stakeholder in the GALILEO Program and is promoting a wide variety of products and applications. On the TLC front, ASI, in collaboration with the Italian MoD is collaborating with France, CNES and DGA, on the bilateral program ATHENA-FIDUS.

3.3 The COSMO-SkyMed Constellation

Since its inception, COSMO-SkyMed, shown in Figure 3.1, has been a Dual-Use *End-to-end* Earth Observation System aimed at establishing a global service able to provide data-products and services relevant to both Civil, Commercial as well Institutional, and Military users.

Figure 3.1 A schematic of the COSMO-SkyMed Spacecraft and its orbital layout.

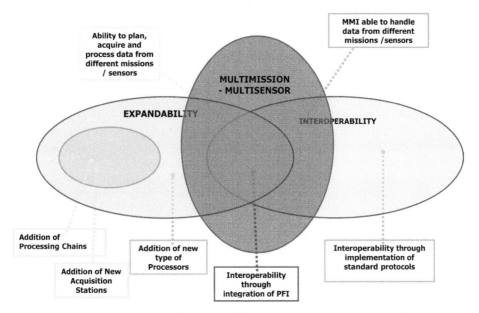

Figure 3.2 The Interoperability-Expandability-Multimission/Multisensor Paradigm
adopted by the COSMO-SkyMed System

International initiatives, such as GMES (Global Monitoring for Environment and Security), GEOSS (Global Earth Observation System of Systems) focus on finding viable ways to provide cost-effective solutions for future Earth Observing systems, granting users with simple access to Multi-Mission/Multi-Sensor (MM/MS) capabilities. MM/MS being the ability to request, process and manage data related to different observation sensors.

The MM/MS is the key point of innovative EO architectures and is achieved by combining Interoperability and Expandability functions to Partner Furnished Items (PFI), in addition to other dedicated MM functions, as deemed appropriate, to address the specific needs of either Defense or Civilian users. PFI items are envisaged to *locally* provide the Partner with capabilities of Mission Programming, Acquisition, Processing and Cataloguing of Sensors data.

The COSMO-SkyMed System Architecture follows these guidelines having pursued, during its development, solutions able to satisfy the whole MM/MS cycle, namely from the users' request throughout the delivery of the multi-sensor products.

The system Expandability is achieved through the capability to integrate and operate the PFI from Partner systems into COSMO-SkyMed architectural elements.

Reciprocally, COSMO-SkyMed User Ground Segment (UGS) can be configured as PFI, to be "exported" towards others Partner EO systems. These PFI are fully defined in terms of software and hardware architecture, operativity, and logistics aspects.

The system Interoperability consists in the capability of exchanging data and information with external heterogeneous systems according to pre-defined agreed modalities and standards, providing access to a variety of EO systems worldwide, mainly for Civilian Institutional, Commercial and Scientific Purposes.

3.4 GALILEO Global Navigation System

GALILEO will consist of a constellation of thirty satellites in Medium Earth Orbit (MEO) distributed on three orbital planes. The satellite geometry has been designed for the launch of multiple satellites with the Ariane launcher while smaller launchers are envisaged for the replacement of individual satellites. GALILEO system will make extended use of sophisticated technology, including highly advanced atomic clocks thus yielding greater accuracy and stability while meeting light weight and low power requirements.

GALILEO is the Civil European **G**lobal **N**avigation **S**atellite **S**ystem that will provide Timing and Positioning services to:

– **Commercial Users** through Open Service, Commercial Service, Safety of Life, Search and Rescue;
– **EU Governmental Users** through Public Regulated Service (PRS) as a robust, undeniable and uninterrupted service for the Public Authorities responsible for Safety of People and National Security.

3.5 ATHENA-FIDUS Telecommunications Satellite

ATHENA-FIDUS is an Italian-French joint mission conceived to meet the needs expressed by the National Institutions and the respective Ministries of Defence.

The Satellite will offer high-capacity multimedia services to be integrated with terrestrial hybrid networks, making great use of advanced technology to optimize its development and the operational costs. The Space Segment is designed to provide full coverage of French and Italian National territories and extra-national "Theatres of Interest".

3.6 Toward an Integrated System: SYNTHESIS

Complex Systems require an innovative approach therefore ASI is planning to adopt a Spiral Development Approach. Spiral Development is a risk-driven process to guide stakeholders through a cyclic approach aimed at incrementally building the degree of definition and implementation.

The definition of the system will start with a set of "Core Requirements" and then will be brought through development by using "Anchor Milestones" in order to ensure stakeholders commitment to mutually satisfactory system solutions.

To this end, ASI has started the program SYNTHESIS an acronym which stands for "Synergies Generated by the Interoperability of Space Infrastructures

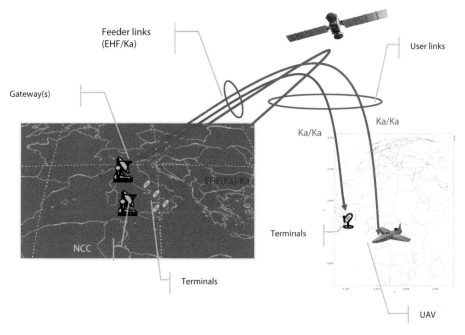

Figure 3.3 A Schematic view of ATHENA-FIDUS System and Services

for Public Regulated Services". SYNTHESIS is a Long-Term program intended to place side by side Earth Observation, Navigation, and Communication Systems in order to provide National and European Institutions with Public Regulated Services (PRS) to support a rapid response to crises, disasters and emergencies.

SYNTHESIS will be based on (A) the National Space Monitoring Center (NSMC) in Italy and (B) the European Space Monitoring Center (ESMC) in Europe.

Therefore the existing "GALILEO NATIONAL ENTITY (GNE)" will be used as the basic asset to build-up the NSMC in order to provide combined Public Regulated Services (Navigation, Communication and Earth-Observation) to accredited National institutional users, based on certified security standard & procedures.

At European level the "GALILEO SECURITY MONITORING CENTER (GSMC)" shall be the basic asset to build-up the ESMC in order to provide combined Public Regulated Services to accredited European institutional users, based on certified security standard & procedures.

SYNTHESIS will be focused on major emergencies and will provide in real-time, to the competent authorities, as shown in Figure 3.4, space Navigation, Communication and Earth-Observation data, on the basis of security requirements.

The pilot application will consist in validating the *GSMC* and *GNE* capability to provide acquisition and regulated distribution of space-borne and airborne observation data, including GPS and GALILEO, to accredited Civil

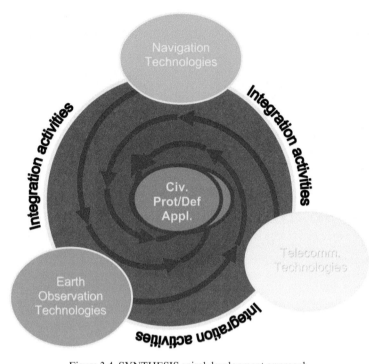

Figure 3.4 SYNTHESIS spiral development approach

Protection Authorities, involved on Surveillance and Monitoring of areas affected by emergencies.

The systems that will be used for the development and test phases (2007–2011) will be: GPS/EGNOS, COSMO-SkyMed and GMES Satellites, airborne platforms and Mobile and Fixed Communication Systems.

3.7 Conclusions

Over several years ASI has been involved in Dual-Use Systems, including related products and applications.

ASI has been developing an infrastructure whose integration offers valuable services to Institutional Users.

SYNTHESIS (Synergies Generated by the Interoperability of Space Infra-structures for Public Regulated Services) is a Long-Term program intended to place side by side Earth Observation, Navigation, and Communication Systems in order to provide National and European Institutions with Public Regulated Services to support a rapid response to crises, disasters and emergencies. The Program is the necessary step for the consolidation of the roles and the functions of the GSMC and the National Entities, which are the two pillars for managing the security of the system.

ASI Proposal aims at involving as many National and European Institutions as possible in order to provide the necessary support and guarantee its sustainability over time.

ASI is ready and willing to collaborate with other Institutions in order to involve them as soon as possible in the process of defining and developing Integrated Systems for Dual Applications.

Chapter 4

Systems-of-Systems and Processes Key to Dual Use

Paul E. Gartz

IEEE Systems Council Operating Unit

4.1 Introduction

Dual Use is not new and has a tremendously humorous aspect: the imperial roman chariots of Rome had a standard developed for the axle distance, that is the length between the two wheels of the chariots, which resulted from the width of the two horses pulling the chariot. Due to the tremendous scope of the roman empire, this standard also moved to what is now the United Kingdom, where it was adopted for the railroad tracks at the time of the steam engine, as that was the standard of the carts (4-foot, eight-and-a-half inch standard), as the ruts on the roads forced the carts to have that axle length, being the cart ruts the holes from the cart wheels. To show the permanence of ancient military standards in the commercial field and the relevance of Dual Use, let's move from the industrial revolution, which modernized transportation with the railroads, to the twentieth century and the Space Shuttle: the two solid state rocket boosters are built by Morton Thiocol, a company which had to be located in the higher mountain regions of north western Utah (USA), from where they had to be transported to the launch site in Florida by rail. As the diameter of the boosters had to meet the railroad limitations, it turns out, eventually, that a design characteristic and constraint of a part of the Space Shuttle — a twentieth century Space exploration system — is determined by the width of the horses setting a standard over 2500 years ago. So, there's nothing new with Dual Use.

There are many types of Dual Use, and many definitions, actually. One of the key definitions of "Dual Use" — not universal, anyway — is "a new system that is being developed from the start with both defense/security and commercial applications". Another typical one is the one that occurred particularly during the 1960's, during the cold war, into the 1970's, which is the use of Space and military technologies (like fuel cells and electronics) to be able to jump start many high value industries (for a value in the range of trillions of US dollars) in the commercial sector, as the ones of the famous Silicon Valley in California. On the other hand, you also have the opposite direction movement of Dual Use, where something is developed originally for commercial purposes and then is re-purposed into applications for defense and security uses: a classic application is the ability to take commercially-developed

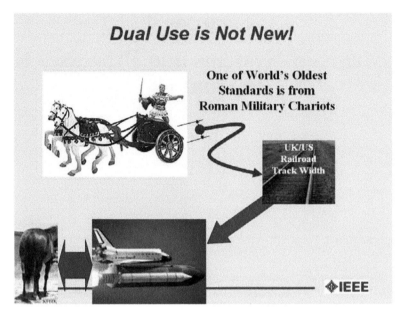

Figure 4.1 Dual use evolution

aircraft and re-purpose them into tankers, cargo and electronic surveillance aircraft like AWACS. A more recent development, for example on the occasion of the most recent wars, as the one in Iraq in 1991, was driven by the urgent need of capabilities that had not been envisioned before going to war: this led to the use of COTS — Commercial Off The Shelf — equipment, such as commercial cars ("dune buggies", sports vehicles designed to go on the sand) to drive in the desert. A fourth category is very interesting and extremely controversial: these are major facilities or equipment developed for presumably peaceful means (commercial satellites, nuclear power plants) which result in capabilities different from the advertised intention. In the case of satellites, this gives rise to very strong political discussion, as they may well be commercial entities, but have spy and intelligence gathering applications: all this is considered "Dual Use". So, as soon as you talk of "Dual Use", because of the nature of the subject, you immediately talk about defense and security.

4.2 History

Taking a look at the history of recent warfare, to see where this entire context fits, there is an interesting pattern occurring in the twentieth century: from the trench warfare of WWI, a war of attrition, we moved to the first truly global war, WWII, where speed became a real critical aspect, and then to the Cold War, WWIII, begun in the late '40s and declared in the '50s (just to quote Sir Winston Churchill, the "Iron Curtain") with consequences of nuclear destruction which we were all incredibly lucky not to face. Now we are into WWIV, where terrorism is worldwide, and this

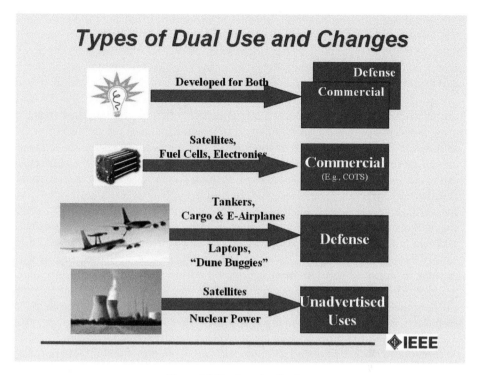

Figure 4.2 Dual use Applications

is again a completely different type of warfare: in fact, each war is a different type of warfare, requiring a different approach in every aspect because of its nature, the problem being in equipping, tactics and training for fighting the war. WWIV is becoming more like WWIII rather than WWII, meaning that what is very key to the approach is not just about the killing and the winning of the physical part of the war, as in WWII, but all the things that surround it: for example, WWWIII was of course much about economics and military armament — "containment warfare" — and what is key to WWIV, also, is that we are now linking security and defense for the first time. In some parts of the world, here in Europe, this has been done for much longer than in other parts of the world, for example in the USA, where there is strong separation between local control of security forces (Police, Fire, First Responders) and the Defense of the country, which deals with external threats. This has very strong boundary separations in the US, for example, where Ministries and Departments are still struggling to figure out how to work together, after six years' discussions, to solve a very difficult problem with a lot of politics inside.

The key of this is that Dual Use sits in a context where we move across the top of spectrum, into this World War where we do not have a really easily defined enemy to fight, not a Nation State but individuals that can come and go when they choose, and attack cultures — for cultural, economic or religious reasons — which may need to be part of a "total security package" to be addressed, and Dual Use comes in this picture.

What is the money context of all this, at macro level? Let's use the US data — US Defense spending versus Gross Domestic Product (GDP) — which represent much of the world spending. For example, when WWWIII was over, after the Soviet Union collapsed, in 1992 the new President of the US Bill Clinton took advantage to gather savings, closing bases all around the world and launching a peacetime reform to capture a "peace dividend", with the implication — in the technical fields — that the relative influence that the US Defense industry (which has a budget today of around 400 billion dollars, still not small by any means) and the military over the acquisition chain and the size of the market levelers (and therefore the influence on the market) has been steadily in decline, both in MIL SPECS and especially in the electronic equipment. These days nothing can be done without electronics anywhere (from iPods to dune buggies, to running a war) as it is a key everywhere: the message, fundamentally, is that once there is a problem in being able to control the markets that are governed by commercial industry around the world, there is a problem in the acquisition chain of what is needed to run these wars. In addition, simply because of the "peace dividend", there was a great restructuring, at least in the US (some of this has been continuing within Europe, as well) and in the management of the acquisition chain, moving to a completely different type of warfare: the sector of Defense in the US at the time — 1994 — was addressed by the Bob Perry Memo, which did many and some of the most dramatic changes ever, including the elimination of military specifications, "the roman chariots", which were removed with a plus and a minus, including some of the consequences that the defense characteristics could not keep up without having enforcing functions of MIL SPECS, with problems on some systems engineering of some of the large US Defense projects. Anyway, the goal of the Bob Perry idea was basically to open up the Defense contracting markets to allow the individual contractors to choose the methods and technologies they would use, as supposed to be enforced to use them under the control of the Department of Defense, with the intention of being able to be using more commercial practices in the actual development process. This did not happen as much as they liked, but it is still an upturn.

The last remark is about the new generation that is coming up: years ago I had the chance to be with Jim Smith, who was the Chief Architect of the flight deck of what was then the F-18, back in 1981, just before it was released. He said: "we've designed the flight deck of this airplane for the kids that are 16, right now, starting to play with video games". Similarly, if we take a look at the generation that is coming up right now — those that will be working in Defense contracting, commercial fields, military services — we see the same characteristics: they are better educated and more aware of technology, their brains are wired by the way they have been growing up, differently from many of us, and these brains are moving faster than ours (quarter seconds instead of minutes for reading), they are pushing all of us — half a generation older than their one — into completely new domains and uses of these things; they are used to have the very best in terms PCs and technology they can get, they are not willing to tackle difficult jobs without the right type of equipment, and — militarily — the result is the useful life of technology is now

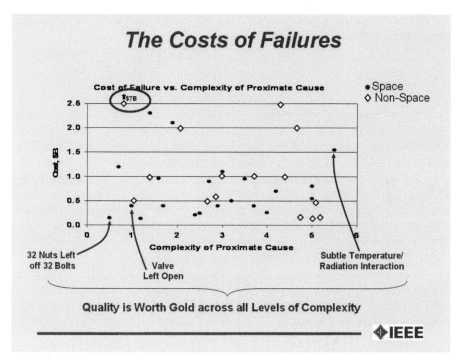

Figure 4.3 Cost of failures in space and non-space programs

changing to be driven by these same people driving the same commercial institutions for different technology lifecycles.

Another major aspect of these things are failures, as they are very expensive: as soon as you go into Space, you have to multiply any of the cost of failures by somewhere between two and three order of magnitudes, up to a thousand or more times higher. To see a real quick example in considering Dual Use, just examine the scatter chart that takes a look of the cost of failures (in billions of dollars) in space and non-space programs with respect to the complexity of the failure that occurs, the intention being to study where failures occur and their relative cost in order to affect the acquisition cycle and the engineering, development, integration and test cycles. Let's first look at the failures in Space programs. A very simple failure, 32 nuts left off 32 bolts, was very cheap: about 200 million dollars; another one, more complex but still extremely simple, that is a valve left open, had a cost of around 450 million dollars; a more complex failure, but well within the range of anything engineering should have taken care, that is subtle variations of temperature and radiation interaction, caused a cost in the range of 1.6 billion dollars. If we now look at non-Space type applications, the scatter representation of all points says it does not matter if it is for Space or non-Space type applications: in any range of complexity of failure, even for low complexity failures, all across the horizontal axis of complexity you can still generate billions of dollars in cost of failure, either you are in Space or on the ground. This is a major fact to be aware of, when developing these things, affecting the choices of Dual Use.

4.3 Systems-of-Systems

Shifting to the Systems-of-Systems context, where are we going? Taking a look at the past, the word first digital commercial airplane, which was certified in July 1982, had a number of subsystems onboard, just the same as a space vehicle, the Space Shuttle and so on. Taking a look at the integration, and moving forward over a couple of decades, systems became more integrated, and fundamentally driven by the digital technologies becoming available, allowing for greater packaging, more processing and more memory, and higher speed networks. Today, what we are moving into, is that the systems that have been integrated are now coming at another level of integration, in the Systems-of-Systems arena: Disaster Management, Communications & Navigation, Defense & Security are some examples.

So, the question is: what is a "System-of-Systems"? This is a relatively new term, which needs to be defined. Like any other system, a "System-of-Systems" is another type of system. But what distinguishes it from just a big system are some characteristics, we identified in the IEEE Systems Council, with many experts in large scale systems integration: independence of components, interoperability of components (as opposed to integration), geographical distribution, evolutionary growth, emergent capability, and unambiguous exchange of information. Fundamentally, a System-of-Systems (SoS) is a new way of looking at the interdependencies of the major systems that we already have today, in this idea of a larger context, conceiving the idea of working together better as a total entity for broader goals, having national or global significance, as for an integrated Transportation System (air, sea, land, and eventually Space), grouping systems in something larger having significance to the people's economy of the countries of the world.

In a traditional engineering development cycle and acquisition scheme, to build a new system the process follows a top-down approach. As with the Systems-of-Systems the systems are already existing, and the issue is to make them work differently, the consequence is that the approach has to be the opposite, that is integrating bottom-up, that is to modify the existing systems along the vision of making them work together to some higher level goals. The difference between top-down and bottom-up approaches in terms of "what to do" is enormous. Quite often there are incredible skills for the top-down approach, the traditional one, whereas the skill base for bottom-up is scarce and needs to be developed.

Now let's see what some of these systems are, in more detail. If we take a look at independent functional capabilities in the world today, we have ability for transportation (air, sea, land, even Space), to observe what the Earth is doing and conclude things about it (to predict how to run businesses or wars, indifferently), to deal with human health, and of course Defense & Security for countries. Each one of these domains is evolving into one of the Systems-of-Systems, as you can get something better out of it, the idea being to optimize systems with interoperability — some of these at the global level, as for transportation and logistics — as this opens potential for benefits and new markets, having a lot of leverage to make a lot of money.

Let's focus on Earth Observation and Defense & Security. If we take a look at the technologies that are going to allow to create Systems-of-Systems out of today's individual systems, a number of them are common to almost all of these systems, or at least some. Sensors, communications, waveforms, materials, information technology, networks, computing and software, information fusion, signal processing: all of these are many of the key technologies that make the glue of Systems-of-Systems, a glue that brings together the puzzle pieces of individual systems to create a new System-of-Systems in each technology domain, each puzzle piece being an IEEE domain.

Different sectors of the economy around the world respond to different motivations and incentives: the classical government (MoD / DoD) approach is to figure out what you are trying to do in your sector, issue a Request For Proposal (RFP), and then issue a contract towards the acquisition cycle. In doing so, there is often a very detailed statement of what is required, and government and Defense are key. A second example, a completely different way of doing business, is moving into the commercial side of COTS, specifically for Dual Use: here you are risking your own money and intelligence, as there is no contract as there was in the first case. The risk is losing money, marketplace, investment and intelligence, as it happened to the US over the past thirty years, losing trillions of dollars in favor of the Japanese market, by not listing the customer base effectively. Third case, the Internet: the speed of moving to market is a key capability to use, which is from two to three or — sometimes — ten times faster each year. Some examples of each of these Systems-of-Systems are: the information battlefield — Systems-of-Systems with network-centric operations, with the foot soldier linked with satellites; GEOSS (Global Earth Observation System-of-Systems) — it is the only one that's truly global already in existence today; Saudi Arabia — probably the single chance opportunity of building something from scratch, thanks to the lot of money available and the lack of infrastructure. Each of these Systems-of-Systems, like the above three, are extremely different ways of doing business, and they are very important because you want to pick and choose from all of them if you are doing Dual Use.

Secondly, a little bit more in this exact subject, if you look at the way you integrate all these pieces for Systems-of-Systems, there are differences between how the commercial and the Defense people do this, worldwide, mainly for the reasons of how the money flows: once you win a contract, you basically do your technical job and deliver a good product; if you are risking your own money, though, you have better get your customer and your business part integrate, business meaning how much money you are really going to make and when, and, if you do not have this integration, doing a trade-off study with different technologies and maybe the entirety of the technical definition, into what is the mission use into the market and how much money you are going to make and when. This is the commercial Systems-of-Systems perspective, very critical and worth a lot of discussion.

4.4 Dual Use Applications

Back to the three examples mentioned, here are some quotes, first for Security & Defense, from Donald Rumsfeld, former Secretary of State of Defense, who had a vision for transformation: "to achieve joint, network-centric, distributed forces capable of rapid decision superiority and massed effects across the battlefield". This is a very important statement, and the keyword is "rapid". Here's another quote, from the war fighter in the field, Colonel John Coleman of US Marine Corps, former Chief of Staff of the First Marine Expeditionary Force in Operation Iraqi Freedom: "I don't want to wait for an 80% solution. What I need is a 20% solution now that gives me a 5% advantage over the enemy". The keyword is "now", which represents the need of having something quick — maybe COTS — as dune buggy vehicles, within a week and not in five years' time. In Europe and through NATO, the same concept of Systems-of-Systems is translated into focus on Expeditionary Operations and Interoperability, Information Sharing and Standards (to be kept at a minimum in order to get the job done), and Co-evolution of Operations and Technology.

The second example, GEOSS, fundamentally means observing the Earth, processing data, figuring out what the Earth is doing and using it commercially or militarily. A simple application is weather and climate: over millennia, nobody goes to war without understanding the weather. Some 4000 years ago, when the Chinese were trying to invade the Japanese, they were turned back by an enormous storm, which sank the enormous amount of their ships, and the Japanese called it "kamikaze", which means "the divine wind". Anyway, as in D-Day and Desert Storm operations, all military forces use existing weather data or have their own, and in the US there are three different Agencies for the Army, the Marines & the Navy, and the Air Force. Climate: some degree temperature differences basically determine which parts of the world are going to be inhabitable or not, and of course there is a lot of research and discussion about this and climate change.

If we take a look at the whole of observing the Earth (the physical, biological, social and economic parameters) and look at what GEOSS is doing, which is affecting commerce, economics, safety, human life and the environment, we see that the effect is amazing, as it touches every single sector (from insurance to financing, from Defense to infrastructure on the Planet) at very expensive levels: just disasters affect somewhere around 30% of GDP of a country that is involved, and for the world this means a risk exposure of 17 trillion dollars a year. GEOSS was started in 2003, in a cooperative way across the world, by Vice Admiral Conrad C. Lautenbacher (former 3-star Admiral in the US) and Deputy Secretary of Commerce, and now (April 2007 data) has 66 Nations (including all Europe) and 44 Non Government Organizations (NGOs), including IEEE — which is a very active NGO member, increasing its role every year — with a 10-year implementation plan in place.

Saudi Arabia is an interesting example. Like many parts of the Middle East, Saudi Arabia have an enormous wealth they have not spent under infrastructure. The first core competence of the country is of course oil but, in case of oil becoming less valuable, they have looked around to seek the potential for a second core competency, concluding it is logistics: what Saudi Arabia are looking at is developing a "land bridge" between the Far East (China and India in particular, over the Himalayan mountains) into Europe, for oil and all kind of materials. They are building six entire new cities, and buildings around the world, with the absolutely staggering level of investment of 1.1 trillion dollars over 10 years, and about 50 billion US dollars per year already spent in the past two years, an enormous opportunity to do this in a leapfrog fashion, faster than China, who is developing similar capabilities and is the largest user already.

Why should we care? Because where money is made will change, the mission is going to change, Defense & Security are changing mainly because of speed, as they must deal with terrorists that do not deal with bureaucracies and buy the very latest technology they can afford, and Systems-of-Systems will emerge as macro-integrators.

Finally, two very interesting potential applications for Dual Use are worth being mentioned, two ".com" tools that are already available right now. The first is Google Earth, who has developed an entire capability on the Google search web site, basically to provide real-time geospatial data, which today are essential in the commercial and in the military domains, driven by the GPS system, a system which is essentially free but is a multi-million-dollar business. The second example is Second Life, a web-based program, basically an animated character: anyone of us can go to the web site and create his/her own character, that is who you would like to be in an alternative reality. The interesting aspect of this is that it is real time dynamic, and the people that are participating in this are interacting dynamically as if they were in a completely alternative universe. How does it work? You use real money in it, from buying clothes to investing in real estates! It turns out that there are about seven millionaires now, who have made money by investing in this alternative universe, just like in the real universe. Anyway, it is a capability to simulate in a simulation environment which is basically free. And these are just two of the many potentials.

Chapter 5

Strategy to Enhance the Technological Aerospace District (DTA) Impact in the Global Dual Scenario

Gerardo Lancia

Filas S.p.A., Roma — Italy

Regione Lazio's Aerospace Sector: Key Indicators and Value Drivers

- **250 prominent sized companies** concentrate in different areas of industrial expertise
- **30,000** employees
- **5 Billion** Euro turnover
- **10 Research centres** (ASI, CNR, CSM, ENEA, ESA/ESRIN, INFN, INAF, CAA, RSV, Centro Atena, ecc.)
- **5 Universities** (La Sapienza, Tor Vergata, RomaTre, Cassino, Viterbo)
- **4 Technological Centres** (Tecnopolo Tiburtino, Tecnopolo di Castel Romano, Pa.L.Mer., Campus biomedico)
- **4 Engineering Faculties, 12 Departments, 30 Postgraduate** and **Graduate courses** and many research projects managed by ASI, ESA, CNR, ENEA, INFN.
- **3000 professors, researchers** and **specialists** involved in R&D activities in aerospace fields.
- **Support services for the technology transfer and start-up creation**: Incubators centres (i.e. ESA incubation centre, Bic Lazio, E2bLab Tor vergata, Sapienza Innovazione, ILO Roma 3, ecc)

The Constitution of the Aerospace Technology District in Lazio (DTA)

- On **June 2004** the **Regione Lazio** signed the **APQ6** (Framework Agreement) with **MEF** (Ministry for Economy and Finance) and **MUR** (Ministry for University and Research) for the constitution of the DTA
- Regione Lazio entrusted **Filas** to realise and coordinate DTA's activities

Resources	Areas of interventions
62 M€ (30 M€ MUR + 32 M€ Regione Lazio) available for the following activities: 1. Industrial Research 2. Education & Training 3. Technology Transfer 4. Support to Innovation Projects 5. Start-up creation 6. Seed and Venture capital for SMEs 7. Development and management of infrastructure and laboratories 8. Large scale demonstration projects	• New planning and design methodologies, innovative materials, new production technologies • New technologies and methodologies for telecommunication and remote sensing space systems and equipment • New technologies and methodologies for avionics systems and equipment • New technologies and methodologies for avionics and land communication systems and equipment • New technologies for air traffic and airport traffic management

FILAS' 5 Main Areas of Action for the Management of DTA

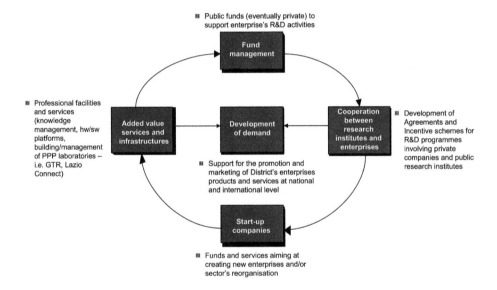

Instruments for Supporting DTA

Fund Management (already launched)

- **MUR's First Call for 2005 DTA proposal (14,05 M€ — closed)**
  Industrial Research, Education & Training in Lazio's aerospace sector
  - *35 proposals evaluated for a total of 99 M€*
  - *9 projects selected and launched with 14,05 M€ public funding*

- **PITT Technology Transfer**
  €100.000 grants for TT between Universities/R&D centres and regional private enterprises
  - *11 TT projects approved in 2006 and 4 in the first semester of 2007*
- **Innovalazio Prize (now open!)**
  Annual regional contest (in its 3rd year) awarding the 3 best performing innovative companies with a €100.000 prize
  - *€600.000 won in 2006 and 2007 and more than 80 quality marks awarded*
  - *closing date for the 2007 edition: 10 October 2007*

Instruments for Supporting DTA

Fund Management (under development)

- **MUR's Second call for DTA proposals (14,6 M€)**
  aimed at supporting Industrial research projects and training in the aerospace sector in the Lazio region
  - *Beneficiaries: Universities, Research Centres, Enterprises*
- **Regional call for proposal in satellite navigation sector for SMEs (2,35 M€)**
  Regional Call for proposals aimed at supporting the development of new products/services and applications based on SATNAV technologies and the future GALILEO system.
  - *Beneficiaries: SMEs, consortia of SMEs*

Instruments for Supporting DTA

Start-up creation

- **Seed and venture capital (grants awarded on a rolling basis)**
  Grants for SMEs as per L.R. 2/85 for innovative initiatives in high tech sectors. Filas' participation in companies' equity together with private investors (i.e. VC)
  - *in 2006, 3 participations for a total value of €1.480.000*
  - *2 Newco dedicated to the development of innovative products for satellite applications*
- **Business Lab Filas (grants awarded on a rolling basis)**
  Grants for high tech start-ups to finance feasibility studies, business planning, legal and financial assistance.
  - *2004 to date: 15 start-up in the Aerospace sector (42% success rate).*

Instruments for Supporting DTA

Added Vale Services and Infrastructures

The GALILEO Test Range (GTR)

The Lazio aerospace industry is a European cluster of excellence, specialising in products and laboratories — i.e. GALILEO Test Range (GTR) — that will serve the GALILEO system, offering optimal conditions for the development of European enterprises in global markets

- *the development of the First Stage of GTR infrastructure is completed and testing is about to start*
- *ASI-Regione Lazio Agreement signed the for the launch of the Second Stage of GTR infrastructure*
- *GTR infrastructure upgrading and opening of laboratories (in experimental phase) in favour of enterprises and research centres for projects research*

Rome's application as candidate city for GALILEO Supervisory Authority's (GSA) seat *Regione Lazio's application is currently highly ranked and well positioned despite strong competition [decision still pending]*

The ENEIDE Soyuz Mission

April 2005: 17 experimental research projects have been carried out on the International Space Station (ISS) in this first regionally sponsored human spaceflight mission

Instruments for Supporting DTA

Cooperation schemes, Enterprises' and Innovation networks

Designed and launched 3 horizontal instruments to foster the cooperation and aggregation processes in the regional Industrial System:

- **Lazio CONNECT**
  A collaborative platform with legal and technical schemes in order to support the creation of a virtual enterprise network for the district
- **DTAwebDB — Web portal for DTA's business community**
  Interactive Portal mapping Lazio's DTA expertises, technologies, products and services
  - *100+ companies registered up to now — Beta version on-line soon*
- **Space2Land — from space technologies to civil industrial applications**
  Initiative aimed at supporting the transfer of space technologies into commercial and civil applications involving Public Demand representatives
  - *14 projects in fields of Infomobility and Transport, Territorial Management and Structure Monitoring (Civil engineering, Cartography), Safety & Environment.*
  - *round tables involving industry, R&D and Demand/Users [ATAC, Trenitalia, ANAS, Ares 118, Protezione Civile, Regional and Municipal Councils]*

Instruments for Supporting DTA

Support for national and international demand: NETWORKS

- **NEREUS — European network of regions using space technologies**
  Regione Lazio participates in the constitution of a European Body that will represent EU Regions interested in space technologies aiming at fostering the development of space technologies for civil applications.
  Main partners: Midi-Pyrénées (France), Baden-Württemberg, Bavaria
- **ENCADRE — Network of European Aerospace Districts**
  Coordinated by EU DG Enterprise, Lazio's DTA participates in the network of European Aerospace districts.
  Main Partners: Cluster Stuttgart (Germany), Cluster Prague (Czech Rep.), Cluster Baviera (Germany)
- **Match-making events with international delegations**
  Participation in Business match-making events (in cooperation with Industry Associations) with delegation of international countries
  Main Countries: Malaysia, Netherlands, Mexico, Canada, Australia

Instruments for Supporting DTA

Support for national and international demand: PROJECTS

- ***SIDEREUS — Project with Asia (EU projects — Asia Invest)***
  The project, coordinated by Filas, will host a business match-making between Lazio's DTA and Asian aerospace enterprises. Event planned in Beijing in 2008. The Asian countries involved are China, that will be the hosting partner, India, Singapore and representative companies from the ASEAN countries. European partners are BavAiria (Germany), GAIA (Spain), Hertfordshire Business Incubation Centre (UK)
- ***CLUNET — Cluster Network***
  This FP6 project, financed under the framework of the Pro-Inno Europe Initiative, aims at identifying and analysing regional clusters and their strategies, policies, and financial schemes. Within CLUNET, Filas is carrying out studies in the Aerospace, together with Life Science and the Audiovisual sectors
- ***Other EU FP7 projects***
  Filas/Centro Atena supports and facilitates the creation of partnerships to the 7th European Union Research Framework Programme

Filas

Regione Lazio

Assessorato allo Sviluppo
Economico, Ricerca,
Innovazione e Turismo

Finanziaria laziale
di sviluppo

Chapter 6

Technology Applications for Evolving Security Needs

G. C. Grasso

Finmeccanica SpA

To compete effectively at international level in high-technology sectors, substantial and targeted investments in research and development are increasingly vital.

Finmeccanica, well aware of this, has made important investments in research and development in recent years: it is one of the top 50 companies worldwide in terms of R&D investments as a percentage of revenues and among the top-ten in the aerospace and defence sector (Figure 6.1).

Finmeccanica is recognised for technological expertise and excellence in helicopters, aircraft, space, defence electronics, energy defence systems, transportation and systems integration (Figure 6.2).

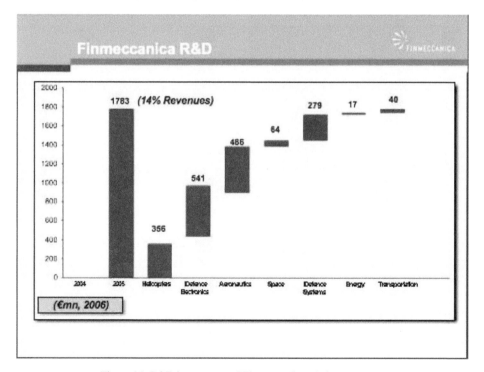

Figure 6.1 R&D investments of Finmeccanica relative to revenue

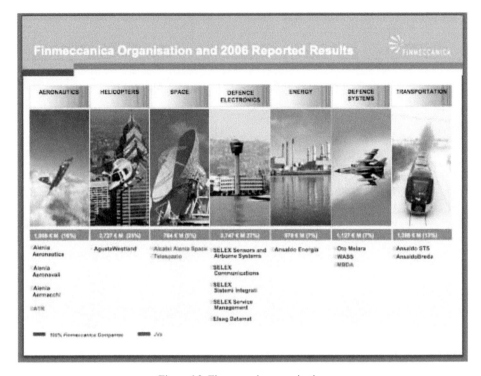

Figure 6.2 Finmeccanica organisation

However, all this is not always sufficient. In the industrial context what really matters is the ability to enhance and sustain over time ones competitive position offering attractive and competitive products and systems which do fulfil and exceed customer requests and expectations and possibly, anticipating his/her needs and wants.

An interesting change is taking place. Once upon a time, an important divide was in place. On one side, those involved with the "military" market, always and only product and services oriented. On the other side, those dealing with the "civil" market, obsessed always and only with the customer. Product/services versus Customers. Nowadays the gap has closed and everybody agrees that what really matters is the customer orientation. Always.

This change has brought further evidence that technology, as such, is neither military nor civil. It is, and always has been, "dual-use". This is very true for the aerospace technologies and their applications.

An important driving force feeding this process of change is the continuous pressure to reduce costs and time-to-market. Today technological developments and relative industrial application are concurrent. The delay between discovery and application is almost gone. Besides the economic pressure, the political and social environment are also playing a major role. Today asymmetric threats and peace enforcing/keeping missions show how thin has become the difference between military and police/rescue forces. The "security" issue has evolved into a need that is neither civil nor military: it is both and it is complex.

In the recent past, threats where simple and with a quite well known origin: land, air, sea…). It was possible to plan the reaction assigning to single specific authorities the responsibility to react, applying the knowledge consequent to the processing of proper data collected by specific sensors' systems.

Today, and even more in the future, a new approach is mandatory. The origin of the threat is unclear. Who should cope with it is no more well defined. Specific sensors are inadequate. A multidisciplinary, multi-responsibility, complex system has to be put in place. Many sensors are needed, massive parallel data processing, fusion and mining has to take place, decision to and how to react has to be taken by different authorities teaming together.

This architecture, known as NCW- Network Centric Warfare, is evolving into the NCO- Network Centric Operation. Homeland Security Systems and Crisis Management Systems must be integrated (Figure 6.3).

It should be stressed that one cannot speak about sensors, without taking into due consideration the respective platforms. The Finmeccanica Group is quite unique being a major player in two of the basic components of integrated homeland security systems: sensors and platforms. Speaking about sensors, Finmeccanica has very significant experience and capabilities in ground based, maritime and airborne radars; advanced avionics, electro-optics, positioning sensors. Its companies master

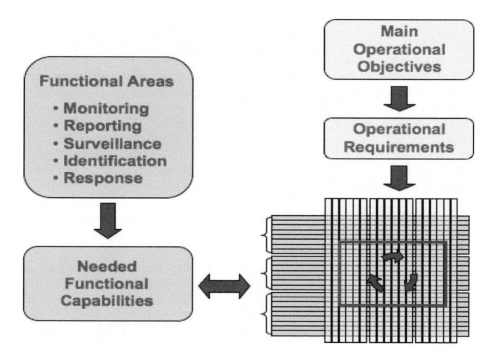

Figure 6.3 Integration of homeland security and crisis management systems

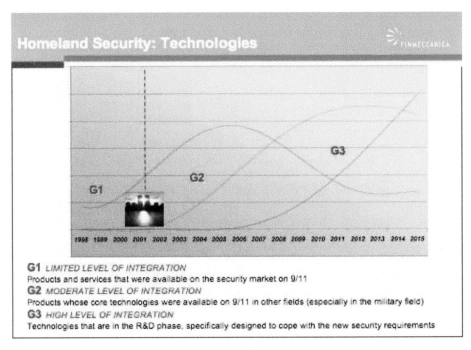

Figure 6.4 Three levels of technological integration for homeland security

the full range of strategic platforms: armoured vehicles; underwater systems, fixed and rotary wing, robotic systems, UAV, ships and satellites.

The concept of dual use technology is also evolving and its boundaries are blurring. No one asks any longer if a technology is or is not dual. The question is: "does the technology enable us to face the threats to the global security and are the resources allocated in such a way that they can be properly managed by different authorities, integrated in a NCO aimed to the global Homeland Security?".

Speaking of Homeland Security, it is possible to classify, taking into account the level of integration, the technologies in three broad sets: G1; G2 and G3 as shown hereby (Figure 6.4).

It is also interesting to give a look at the difference in investments in Europe and in the Unites for "security" products, taking into account the size of the global security market which is in the range of US $55 billion (Figure 6.5).

The investments in Europe are much smaller that the United States ones. In the US the security market in 2006 was worth of US $29,2 billion (0,22 percent of the Gross National Product), whilst Europe spent US $11,4 billion. It seems that one of the consequences of the 9.11 dramatic event, is that the American society assign to "security" a much higher value compared to the one assigned to it by the Europeans. This in spite of the fact that, according to Euro Barometer Survey, April 2007, European citizens assign a top priority to the security issue.

GDP – USA : **$ 13.262 B**

GDP – EU : **$ 14.206 B**

Figure 6.5 Investment in security products.

To provide Europe and its Member States with the needed Homeland Security capabilities many efforts are going-on at the EU level to support new organisational solutions and procedures both at national and at EU level, in order to enable:

- the Dissemination, over secure networks, of timely and relevant information across organizations and Member States
- a Common Operational Picture created by national and regional Control Centers integrating the appropriate data sources and providing situational awareness at the operational level
- the Sharing of existing means and information for co-ordination, co-operation and synchronization across Member States
- the Introduction of new functions and capabilities taking advantage of new, dual-use technologies, sensors and platforms.

The request for "security" dominates our daily life. The security needs and wants will continue to put increasing pressure on the technological development in order to fulfil the requirements imposed by the rapidly evolving threats.

Finmeccanica is well positioned to play a major role in this new, highly competitive and challenging market. Its high tech assets, strong technogical governance, and longlasting experiences in dual-use technologies' applications, allows Finmeccanica to sustain over time its organic and external international growth, continuously and virtuously innovating and creating wealth.

Whenever it will be needed, enabling technologies will be acquired. The existing available wide European technological base enables Finmeccanica to be competitive worldwide, always optimizing time to market, costs, efficiency and effectiveness of proposed products, services and processes. Never forgetting the full respect of citizen's privacy and rights.

Chapter 7

Dual Use in GMES: Analysis of Strategy and Developments

Enrico Saggese

Finmeccanica S.p.A.

7.1 Introduction

The technological revolution which has picked up a significant momentum since the 70's and the 80's with the computing technology, has developed a new practise for that commercial innovations are now quite often driving military modernization. Since nearly all advanced technologies have civil and defence applications, the concept of dual-use technologies is losing its relevance.

This statement seems particularly pertinent with regard to satellite remote sensing.

Indeed, when looking at the marketing policies of commercial remote sensing firms like Spot Image (France), Radarsat International (Canada) or remote sensing businesses in the US, it becomes evident that these companies are somehow plainly offering to enhance, or even create, the imagery intelligence capabilities of prospective clients.

Satellite imagery, and the information derived from it, is prone to a wide range of both civilian and defence applications.

It can be a "force multiplier" for the peace-keeper just as it can for the peace-enforcer or the peace-maker.

Within this chapter we would like to draw attention how remote sensing dual uses are being practised, and which policies these practices imply, particularly in view of the GMES - Global Monitoring for Environment and Security i. e. the European System of Systems for Earth Observation.

By its global nature GMES implies an integration with other space based system, such as for instance a global telecommunication relaying satellite system as well as a space based global navigation system, to fully respond to the real time needs for emergency and security.

7.2 Dual-Use Past Policies and Practices

Remote sensing dual-use issues have traditionally been treated in terms of factual employment of civil space-borne observation systems for defence and, more generally, security purposes.

There is general evidence, however, that commercial remote sensing technology applications are clearly emerging from a government-funded, defence heritage.

This was already the case with LANDSAT, the civilian US land remote sensing system set up by NASA in 1972 which drew on technologies developed under US reconnaissance satellite programs like Corona and the subsequent "keyhole" series.

Commercial high-resolution systems are only slightly modified civilian replica of earlier US defence systems generally developed by the same companies which are now parties to the commercial remote sensing consortia.

Remote sensing technology transfer can also work in the opposite direction. The French SPOT system, e.g., although established as a civilian enterprise, from the outset was also planned to serve as a testbed for a later defence system, i. e. Helios. Not surprisingly, therefore, both systems have a number of commonalities.

Since the end of the Cold War restrictions on the development and operation of high-performance non-military remote sensing systems have been considerably lessened. Firms are at liberty to devise remote sensing technologies with a double use potential. Defence programs are being realized with an eye on possible commercial applications and vice-versa civilian programs are established having defence customers as anchor tenants.

The dual-use notion, therefore, is not relatable to the nature of a specific technology but to circumstantial utilisation and prevailing policy assessment, especially under proliferation policy aspects.

The multiple usage possibilities of modern remote sensing technologies are enhanced by the growing accessibility to high-resolution satellite imagery data. In the same way as with technology transfer, the accessibility operates in both directions.

With the declassification of American and Russian spy satellite archives at least some secret data sets are now available to the public sector.

This development will not lead to full civilian co-use of essential defence remote sensing systems, but the data merging of certain defence and civilian observation satellite constellations is bound to gradually increase mutual accessibility.

This tendency to associate formerly distinctive defence and civil observation assets except, of course, those of a uniquely military character, is continuing to develop and the GMES European project is fully going in this direction.

The process is triggered by forceful motivations, namely to economise investment funds, to ease the operational cost burden, and to gather the private sector's technological innovations.

The defence, in turn, increasingly seek access to the services and products of civil, commercially operated remote sensing systems.

Especially since the Gulf War it has become a habit, if not a routine, to also rely on open observation sources like SPOT or LANDSAT for defence purposes of all kinds, ranging from reconnaissance over targeting to damage assessment.

The usefulness of privately rendered remote sensing services has proven to be so great that defence authorities, both in the U.S. and in Europe, are now contracting firm arrangements with commercial data suppliers.

Formerly under the control of two space powers, namely the U.S. and the Soviet Union, observation satellites in the meantime are becoming available to a steadily growing number of countries. The lowering of secrecy restrictions by the U.S. and Russian governments were deliberate policy acts.

7.3 GMES, the European Policy

The Global Monitoring for Environment and Security is a *global integrated system* for which Space Assets and Services represent the backbone (EO satellite technologies and applications complemented and integrated with satellite communications and navigation), but in-situ and real time airborne data are fundamental complements.

Even if the GMES concept was conceived in 1998 ("Baveno Manifest"), the real start-up of the programme occurred in 2001 through the endorsement by the European Commission and the ESA Councils, then in its Communication to the EU Gothenburg Summit (June 2001), the Commission called for establishing a GMES European Capacity by 2008.

GMES strategy in the development of the EU's role as a global actor has been outlined in a February 2004 Communication of the European Commission, which also identified the major EU policies to be addressed by GMES services.

These can be summarised as follows:

– Europe's environmental commitments, within EU territory and globally, by contributing to the formulation, implementation and verification of the Community environmental policies, national regulations and international conventions;
– other EU policy areas such as agriculture, regional development, fisheries, transport, external relations with respect to the integration of the environmental dimension in the respective domains;
– Common Foreign and Security Policy (CFSP), including the European Security and Defence Policy (ESDP);
– other policies relevant to European citizens' security at Community and national levels, notably the potential that exists for application to, e.g., policies related to Justice and Home Affairs activities of the European Union, such as border surveillance.

The dual use policy is clearly visible.

Looking at the details of the European policy for the GMES development we note that The European Commission has set out a strategy for delivering GMES according to the Council mandate.

This will optimise planned European space and in-situ infrastructure and fill identified gaps to respond to the requirements of service users. Decisions already taken start the process of securing the availability of the space component, the development of which has to be funded by ESA (the "Sentinels" satellites) while the EU and/or member States will take care of the system implementation.

EUMETSAT will in parallel enhance its meteorological and climatologic infrastructures and services.

In addition, the EU Security Strategy has highlighted that Europe faces constantly evolving threats which are more diverse, less visible and less predictable. The Commission has identified security of EU citizens as one of the three main objectives in its work programme. To tackle these constantly evolving threats requires a mixture of civilian and defence solutions. Space assets provide a significant contribution to this goal.

The EU approach to crisis management emphasises the synergy between civilian and military actors. Space system needs for planning and conducting civilian and military Crisis Management Operations overlap.

The Member States in the Council have identified Europe's generic space system needs for military operations and stressed the necessary interoperability between civilian and military users. Military capability will continue within the remit of Member States.

This should not prevent them from achieving the best level of capability, within limits acceptable to their national sovereignty and essential security interests.

Sharing and pooling the resources of European civilian and defence space programmes, drawing on multiple use technology and common standards, would allow more cost-effective solutions.

The GMES programme is currently in the implementation phase (2004–2008). It has now reached a stage of maturity in which it can progress to its development and implementation, with service provision for different areas, becoming operational in a phased approach.

It remains within ESA as an optional programme thus being funded by the member States willing to improve their possibility to access Earth Observation data.

From the experience gained during the initial period, there is significant support to validate the further development of GMES, to achieve a "core capacity": the initial set of services and the supporting components required to deliver these services on an operational basis by 2008. The capacity has to be built-up progressively, based on clearly identified priorities and by using existing elements whenever possible.

The development of a strong European GMES capacity requires the mobilisation of expertise and competence in industry, research institutes, academic sectors and governmental organisations.

The target of the current development phase is to build a strong user base for GMES information services. This implies that needs are identified and updated and that the services should be reliable and effective.

The GMES scenario is rather complex: it requires the integration of data from space-based and in-situ (airborne, water-based and ground-based) sensors into user-driven operational application services. This approach will be developed in steps through the introduction of pilot phase services. The gradual introduction of services will be based on their maturity.

The challenges are to achieve increased interoperability of systems, to harmonise the standardisation of data structures and of data sharing, to combine data from different sources at different levels and to provide sustainable services.

To meet the goal of operational services by 2008, in the first step, the European Commission judged that three fast track services were ready to satisfy these criteria and proceeded to a pilot operational phase.

The three services already in progress are: Emergency Response, Land Monitoring and Marine Core Services.

The general objective remains to develop an extended range of GMES services which meet user requirements. For this reason, further pilots services have been introduced: Atmosphere monitoring and Security (i.e. external border surveillance) of which some Member States, including Italy, have emphasised the relevance: GMES is the opportunity for a rapid deployment of operational services and the need and the strategic value of the "S" component are confirmed.

An issue related to the success of the GMES service components is the development of a European Spatial Data Infrastructure (ESDI) as foreseen by an INSPIRE directive.

GMES Data policy will need to be developed, to ease acquisition and exploitation by both service providers and users. This approach is coherent with the key objective of INSPIRE to make more and better spatial data available for Community policy-making and to implement Community policies in the Member States at all levels.

The security services should be operational in the very short term, in order to provide an adequate and timely response to the urgent requirements of the European Security Strategy.

GMES services ultimately shall be derived from different space infrastructures which have been built in some cases for other purposes than the GMES only use, and this particular approach leads to complex problems in the governance of the system as well as in the investor rationale.

The "federation approach" of different space based systems is the key to proceed with GMES as a "System of Systems".

7.4 Italy and GMES

Italy, having excellent competences, large industrial capabilities, unique infrastructures, and a strong user community, is one of the European countries that are playing a fundamental role in the GMES program. Furthermore, Italy has undertaken a strong commitment to contributing to the program, also through the GMES Space component.

Italy will contribute in a unique manner to GMES through assets already developed at national, regional and local levels and industrial capabilities that are able to cover all the value chain in the different domains: space segment for earth observation, telecommunication and navigation spanning instruments from microwave to the visible and infrared, data supply and customized services developed and validated in collaboration with the end-user, security rules and standards to answer to commercial and defence needs.

One of the pillars of GMES is the system COSMO-SkyMed, funded by ASI, the Italian Space Agency, and the national MoD, as part of a larger Franco-Italian programme including both optical and radar sensors: ORFEO.

The COSMO-SkyMed Constellation is made up of a space segment with high resolution SAR imagery instruments and a Ground Segment for control and data exploitation both driven by leading requirements: dual use and interoperability.

The space segment, made up of four satellites, is equipped with a multi-mode SAR instrument and is capable of operating in all visibility conditions at high resolution and in real time.

The instrument is fully programmable and handles a wide frequency spectrum. The technology used is state of the art and it is the precursor of the new generation radar that will allow more information to be extracted from the return signal as requested by users for a wide area of application.

The satellites are capable of attitude manoeuvres allowing the acquisition of data from targets on the left and right side of the satellite track.

COSMO-SkyMed provides a few hours revisit time on the same target and supplies worldwide services for both civil and defence needs.

The COSMO-SkyMed program will also embody the facilities distributed over the Italian territory for defence and civil applications located respectively near Rome, in the south of Italy at Matera and at Fucino (Satellite Control Centre and Centre for Mission Planning and Control).

7.5 Conclusions

The GMES scenario is not already frozen and the evolutionary approach is a key to make of this large program a modular evolving system.

Thus additional features could be envisaged for their addition in the near term future; such features are coming from other space based systems capable to enhance the GMES data provision. In the field of Security for instance a space based radar systems able to track moving objects on ground would enhance the possibility for GMES users to monitor land and sea scenarios in case of emergency or critical events.

The Global Monitoring for Environment and Security System is by nature a global integrated system for which present, planned and still to be defined Space Assets represent the backbone, therefore by conceiving such an evolving system it is mandatory to conceive and build complement and integrated Space Assets in navigation and wideband relaying telecommunication, to allow global data transmission in secure and proprietary manner.

Chapter 8
Extending the Internet into Space

S. Venturi*, P. Massafra*, K.-P. Doerpelkus**

*Cisco Systems, Italy
**Cisco Systems, Global Government Solutions Group

8.1 A Space Vision from a Networking Company

The government sector is transforming by adapting the concepts of "Network-Centric Operation". Satellites as an important part of the global communication infrastructure are migrating away from proprietary technologies to architectures which are based on open standards such as IP. Through their CLEO experiment, Cisco has demonstrated the viability of using their commercial internetworking hard- and software in LEO. In a second step the portability of Cisco's router code to space qualified hardware was proven. This will eventually be the direction for future satellite communication projects such as IRIS.

One day, each and every manned and unmanned spacecraft, high altitude platform, unmanned aerial vehicle, or airframe will be a node on the network. The Internet will extend into space by networking all these resources via optical or radio intersatellite links through autonomous routing in space. Terrestrial and space communications will be indistinguishable and will provide access to all kind of services and data anytime, anywhere, and to anybody (Figure 8.1). [1]

The Internet Protocol (IP) and related open standards as defined by the Internet Engineering Task Force (IETF) and other standardization bodies like IEEE are driving the convergence of terrestrial networks toward scalable, flexible and cost efficient network architectures. Operators of fixed and mobile networks as well as enterprises are increasingly migrating from deploying, managing, and maintaining many services-specific networks to delivering all services on a single IP-based network to meet their customers demand for voice, video, and data access anytime, anywhere, over any device.

Governments around the globe are taking advantage of this convergence by migrating their IT systems to an IP-based platform in order to deliver services more effectively and efficiently to their constituents, citizens, and business partners. Network-Centric Operations (NCO) form the foundation for virtualized capabilities and services that enable the defense agencies to use the network to its full advantage. Finally, IP-based networking is becoming more and more popular in the space market, since agencies and systems integrators can leverage the technology and expertise already available for terrestrial networks and reuse it in the space environment.

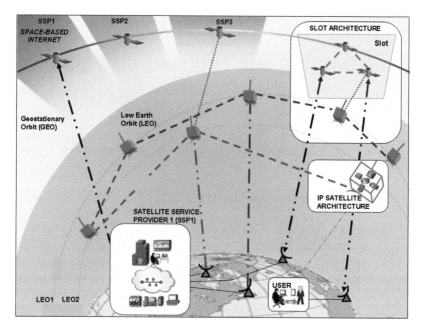

Figure 8.1 Cisco ground space merged architecture

Systems based on "Commercial Off The Shelf" (COTS) or "Dual-Use" technologies help agencies and systems integrators build more flexible, adaptable, interoperable and extensible systems, reducing the dependence on highly specialized hardware and software that impedes the rapid deployment and enhancement of space missions.

The use of IP onboard satellites and space crafts is not a recent story and dates back to the 1970s. The networking aspects of many of these IP-based missions are covered in detail in [2], just to mention a few examples:

- The UK Defence Evaluation Research Agency (DERA) launched the Science and Technology Research Vehicle (STRV) satellite on 17 June 1994. IP software was later uploaded and tested.
- SSTL's UoSAT-12 (University of Surrey) minisatellite was launched on 21 April 1999. An IP stack was later uploaded, giving IP comms in parallel with the AX.25 comms already in use. A Cisco 1601 router was used in the ground station. This was later upgraded to the Cisco 2621 currently used in all DMC ground stations.
- NASA developed the Orbital Communications Adapter, which allows the crews of the shuttle and ISS to use IP. This was used with Cisco Softphone experiments on Atlantis in 2001.
- Communications And Navigation Demonstration On Shuttle (CANDOS) was a payload in the shuttle bay communicating via TDRSS (NASA Tracking and Data Relay Satellite System, which backhauls LEO to ground via GEO bent-pipe satellites). CANDOS used Mobile IP to handle TDRSS

handovers and maintain ongoing file transfers. CANDOS was lost with the seven astronauts aboard shuttle Columbia on re-entry on 1 February 2003.

- CFESat, the Cibola Flight Experiment Satellite, examines radio spectra for ionospheric and lightning studies, using field-programmable gate arrays (FPGAs). As well as science observation, the mission aims to show use of reconfigurable FPGAs to work in the radiation environment of LEO. The satellite payloads were built by Los Alamos National Laboratory.
- MidSTAR-1 was built by the US Naval Academy. It carries the Internet communications satellite (ICSat) experiment.

On 27 September 2003, for the first time in the history of space missions, a commercial Cisco router was launched into LEO as a secondary experimental payload onboard a Disaster Monitoring Constellation (DMC) satellite built by Surrey Satellite Technology Ltd (SSTL) [3]. Today this router still works and demonstrates that Cisco's routing software, the "Internetworking Operating System" (IOS) that powers more than 80% of the terrestrial Internet, can be successfully deployed into space.

8.2 The Evolution of the Space Industry

Historically, global space communications has relied on terrestrial-based technologies, driven primarily by the governments. This model consisted of large, expensive, point-specific "stovepipe" programs with integrators developing their own proprietary solutions and systems based on proprietary protocols, highly specialized hardware and software. This again resulted in long lead times for technology development and system designs that were not durable and reusable.

Today's space community demands faster time to orbit, affecting dramatically the decision-making framework for systems integrators working in the space market. As a result, the pressure to use COTS components as much as possible, rather than highly specialized (and very proprietary) space equipment, and the adoption of IEEE/IETF standards has intensified. COTS products and technologies help to create space systems that are more adaptable and can take the advantage of commercial technology improvements and applications. COTS procurement also enables space agencies to build high performance systems at a lower unit cost, while lowering their total cost of ownership (TCO). NASA and U.S. Government estimate COTS technologies will save up to 50% in overall mission costs, which includes acquisition and life cycle.

The term "COTS" was defined about 15 years ago within the U.S. Department of Defense (DoD). According to the COTS Journal [4], COTS is generally defined for technology, goods and services that use commercial business practices and specifications, which are not developed under government funding and are offered for sale to the general market. Without COTS usage, space agencies and systems integrators suffer from a strong dependence on specialized hardware and software, which prevent their ability to be more efficient, agile and adaptive. For example:

- The use of non standard hardware and software makes interoperability of different missions very difficult and expensive to achieve.
- Cycle times of 5 to 10 years (and sometimes more) for design and development of space missions mean that these are technologically out of date by the time the equipment is launched into space. The technology evolution in the commercial communications industry is very much driven by Moore's law [5] resulting in technology and product cycles much shorter than typical development (and operation) cycles in the space industry.
- Highly specialized equipment drives up development and procurement costs because the space agencies bear the whole amount of development costs.
- Higher support and training costs increase the total cost of ownership over the life of equipment.
- Space agencies are unable to take advantage of technology improvements, new applications developed for commercial implementations and the rapid decline in the cost of standardized components and systems.
- Dependence on specialized equipment drives up the cost of establishing centralized depots for spares.

On the contrary COTS equipments do not have those drawbacks and their reliability has improved so much that the justification for expensive space specific equipment is no longer there. It is however understood that COTS can't fit everywhere since it is not designed for extreme environments like those exposed to high radiation doses.

Based on the considerations above, choosing the right COTS vendor has become a critical factor for both space agencies and systems integrators. Since space contracts must be supported for many years, far longer than normally required in commercial markets, a viable COTS vendor must demonstrate at least the following strengths:

- Ability to supply spares for as long as necessary.
- Compatibility with previous generation of equipment.
- Ability to adapt COTS products to offer special features tailored to the customer's specific requirements.
- Participation in industry and government security organizations.

For the not so distant future Cisco sees a clear trend that the space market will transition from a technology driven to a customer and application driven marketplace. The use of COTS products and technologies which are based on open standards will be instrumental for this transition and a substantial reduction of cost while at the same time development cycles will be reduced. Near-term markets are expanding to include many players such as worldwide, independent, regional and narrowband satellite service providers, commercial avionics manufacturers, civilian organizations and the defense industry. Public and private partnership models will replace business models which historically were driven very much by governments.

8.3 A Cisco Router in Low Earth Orbit

In response to these developments, Cisco has organized a team of senior leaders from the defense, space and homeland security sectors to help educate, evangelized and integrate network centric technology into systems and processes of these markets. The Global Defense, Space and Security Group (GGSG) has been chartered to work with the defense, space and homeland security communities to drive awareness and education around the latest advances in networking technologies. On the other hand GGSG works closely with Cisco's R&D groups to tailor cost effective COTS products and solutions based on customer needs and requirements, mainly from the defense sector, but also from civil agencies and commercial satellite operators.

Surrey Satellite Technology Ltd. (SSTL) (http://www.sstl.co.uk) of Guildford, England, is a major provider of small satellites and related technologies, e.g. the Disaster Monitoring Constellation (DMC). DMC is the first earth observation constellation of 5 low cost small satellites providing daily images for applications including global disaster monitoring, for more information [6], [7]. In a co-operation between SSTL and Cisco a Mobile Access Router MAR3251 was integrated as a secondary experimental payload onboard the DMC Low Earth Orbit satellite. This project, called CLEO (Cisco router in Low Earth Orbit) marked the first time that a commercial Cisco hardware was taken to and tested into space.

The CLEO deployed onboard the DMC satellite consists of 2 PC104 based circuit boards: the Cisco 3251 MAR processor card and the 4 port serial mobile interface card (SMIC). Those serial interfaces are used to connect the other satellite payloads (Figure 8.2). Cisco didn't do any unique software or hardware design. Cisco's commercial IOS software (Release 12.2(11)YQ) was used unmodified on the router, no radiation hardening was done at all. In order to adapt the router to the space environment SSTL performed minor modifications of the router cards:

- Lead holder was used instead of tin solder to avoid "tin whiskers".
- Wet capacitors with vents that would leak in a vacuum were replaced with dry capacitors.
- The clock battery was removed (the time can be set with the Network Timing Protocol NTP).
- Connections between router cards and the motherboards were directly soldered, replacing plastic sockets to resist launch vibrations and thermal cycling.

The router assembly successfully passed a full system flight level qualification testing on its first attempt that includes a temperature range of -35 to $+60°$C and a vacuum of less than 1×10^{-3} Pa.

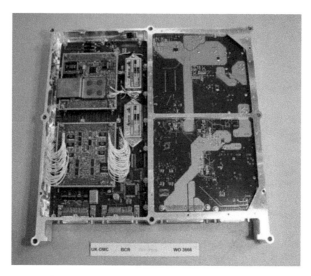

Figure 8.2 CLEO assembly mounted in a rack tray

Although the total power consumption of the combined unit is far less than traditional terrestrial routers (10W at 5V), it's still a significant proportion of the DMC
satellite's available 30W power budget. For this reason CLEO is powered off when
not being tested in order to conserve available satellite power and battery life. Testing
of the CLEO router continues only when the DMC satellite is not otherwise tasked
with its primary imaging mission. More detailed information about the router integration and its operation can be found here [8].

The DMC with its CLEO payload together with a couple of other small satellites was launched from Plesetsk in Siberia on an affordable Russian Kosmos-3M
launcher on September 27th, 2003. Starting in 2004 the router was tested in a couple
of scenarios which are published in a couple of papers, see e.g. [8], [9] and [10].

The IOS image which runs on the CLEO router contains beside the basic IPv4
feature set, Mobile IP, the IP Security Protocol (IPSec), but also IPv6.This newer
version of IP is intended to eventually replace today's terrestrially used IPv4, as
the larger address space and simpler routing tables of IPv6 solve the most pressing
problems with the scalability of IPv4. There are strong requirements for IPv6 from
the US DoD and some MoDs around the globe. IPSec is the common, commercial
technology to secure network assets terrestrially, so it makes sense to reuse this
technology for space environment where appropriate. CLEO was configured for
IPv6 and IPSec use in March 2007, and successfully tested with both features on
29 march 2007 (Figure 8.3), to our knowledge the first test of IPv6 in space ever [11]!

8.4 The Next Step: Embedded Routing

The CLEO router has been in space for over 4 years and has been tested in orbit for
over 3 years. CLEO has been powered up nearly hundred times for testing during

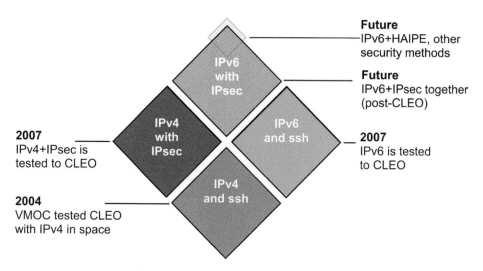

Figure 8.3 Networking features tested on CLEO

passes over ground stations. Today this commercial, non hardened computing device using non qualified space components is still operational in LEO, despite a non friendly environment. CLEO demonstrated the viability of using commercial router code in space, but Cisco does not propose to use their commercial hardware (which is not radiation hardened) for applications in higher orbits or deep space.

As a proof of concept engineers from Cisco GGSG ported commercial IOS code to a fully space qualified radiation hardened processor, the RAD750 from BAE Systems (a PowerPC architecture). The result of that experiment was:

- The IOS porting to the RAD750 was economically doable and didn't cost too much effort.
- The initial performance of the ported code beat the requirements from a traffic model for a LEO sensor which was jointly developed with NASA.

This experiment paved the road for the direction to the future of space- and IP-based communications: embedding commercial terrestrial router code into communications payloads on satellites and other spacecrafts by porting this code to appropriate hardware.

Cisco is participating in a project of the US DoD to test "Internet Routing in Space" (IRIS) which will be managed by Intelsat General. In this 3 year program, Cisco provides its commercial IOS routing code, SEAKR Engineering Inc. of Denver, CO, will manufacture the space hardened router hardware, based on PowerPC G4 processor, and integrate it into the IRIS payload. The satellite — Intelsat 14, manufactured by Space Systems/Loral — is set for launch in the first quarter of 2009 and will be placed in geostationary orbit at 45 degrees west longitude, covering Europe, Africa and the Americas. [12]

Representing the next generation of space based communication, IRIS will serve as a routing engine in the sky, interconnecting C and Ku band transponders, allowing communication between different frequencies without having to switch between them on the ground. The use of onboard switching can later lead to the use of onboard routing functionality with intersatellite links.

8.5 Conclusions

The space market is changing and transforming through the use of COTS technologies, open standards and the standardization on the Internet Protocol IP as underlying communications platform. Extending internetworking technologies and paradigms into space will drive new capabilities and behaviors, shorten time-to-market and reduce costs. For over four years now the Cisco Low Earth Orbit (CLEO) router experiment has demonstrated the viability of COTS networking equipment and commercial software in space. Cisco has proven the portability of their commercial Internetworking Operating System (IOS) to space-qualified processor platforms. The evolution of the CLEO project, IRIS (IP Routing in Space), will further extend the application of internetworking technologies into space-qualified platforms.

References

[1]   Floreani, D., and Wood, L., "Internet to Orbit," *Cisco Systems Packet Magazine*, Vol. 17, No. 3, third quarter 2005, pp. 19–23.

[2]   Hogie, K., Criscuolo E., and R. Parise, "Using standard Internet Protocols and applications in space," *Computer Networks*, special issue on Interplanetary Internet, Vol. 47, No. 5, April 2005, pp. 603–650.

[3]   Waltner, C. "Space No Longer Final Frontier for Cisco Internet Gear," Cisco press release, 26 September 2003. (http://newsroom.cisco.com/dlls/ts_092603.html)

[4]   http://www.cotsjournalonline.com

[5]   http://en.wikipedia.org/wiki/Moore's_law

[6]   http://www.sstl.co.uk/index.php?loc=120

[7]   http://en.wikipedia.org/wiki/Disaster_Monitoring_Constellation

[8]   Wood, L., da Silva Curiel, A., Ivancic, W., Hodgson, D., Shell, D., Jackson C., and Stewart, D. "Operating a terrestrial Internet router onboard and alongside a small satellite," conference paper B-05-03, published in *Selected Proceedings of the 56th International Astronautical Congress*, Fukuoka, Japan, 17–21 October 2005, Acta Astronautica, Vol. 59, No. 1–5, July–September 2006, pp. 124–131.

[9]   Wood, L., Shell, D., Ivancic, W., Conner, B., Miller, E., Stewart D., and Hodgson, D., "CLEO and VMOC: Enabling Warfighters to Task Space Payloads," unclassified track, *IEEE Milcom 2005*, Atlantic City, New Jersey, 17–20 October 2005.

[10] Conner, B. P., Dikeman, L., Osweiler, V., Schoenfelt, D., Groves, S., Paulsen, P. E., Ivancic, W., Walke J., and Miller, E., "Bringing Space Capabilities to the Warfighter: Virtual Mission Operations Center (VMOC)," paper SSC04-II-7, *18th Annual AIAA/USU Conference on Small Satellites*, Logan, Utah, 9–12 August 2004.

[11] Wood, L., Ivancic, W., Stewart, D., Northam, J., Jackson, C., and da Silva Curiel, A., "IPv6 and IPsec on a Satellite in Space," *58th International Astronautical Congress*, Hyderabad, India, September 2007.

[12] http://www.intelsatgeneral.com/pdf/en/aboutus/releases/2007-4-11-IRIS.pdf
      General information about Cisco's space activities can be found at http://www.cisco.com/go/space

PART 2

Dual Use Technologies

Chapter 9

Assets and Technologies for Space-based Dual Use Systems

Ing. Carlo Alberto Penazzi

Thales Alenia Space Italia S.p.A.

9.1 Introduction

After the tragic events of September 11 2001, the national governments have acquired a progressively increasing consciousness of Security, for the necessity to guarantee the safety to their citizens and the integrity of strategic infrastructures.

Nowadays, potential threats are often hidden, less predictable, and more lethal than before, being connected to regional conflicts, organised crime, and terrorism.

In this framework, the Italian initiatives and commitments on Security concern several aspects, from the development of new technologically advanced systems to the promotion of multi-national partnerships with Defense and institutional entities in Europe.

Indeed, the evolution of the scenarii in the European foreign politics, concerning Defense and homeland Security, requires an increasing level of cooperation among the European Countries, and their institutional entities (e.g. Police Forces, civilian protection, regional institutions), as well as common efforts, at least at European level.

These new political scenarii of cooperation for Defense and homeland security require state-of-art technologies and global systems of surveillance and protection.

The Italian institutions and the Defense, together with the Italian Space Industry, significantly contribute in Europe and on a worldwide scale in responding to this technological challenge, through integrated systems that provide services and infrastructures for environmental monitoring and global security, aimed at dual-use from Defense and Civilian entities.

In this perspective, Italy is providing a fundamental step ahead towards the dual-use applications, through the development, launch and deployment in orbit of COSMO-SkyMed, the first constellation of satellites for Earth Observation for dual-use at world, with technologies, versatility and operative characteristics largely superior to any other space-based observation systems so far available or foreseen in the near future.

These Earth Observation systems cover an important part of the space-based capabilities for dual-use. Other space systems, devoted to Telecommunication, Meteorology, and Navigation Support can effectively complement and complete the integrated framework of heterogeneous space assets for dual-use, conceived for environmental monitoring, surveillance of territories, safety of citizens, and security of strategic infrastructures and services.

In this chapter, the key characteristics of these space systems are delineated, along with the specific contributions that Thales Alenia Space Italia is providing to this challenge.

9.2 The Dual-Use concept for Space Systems

The basic idea of a dual-use space system designates its suitability to be simultaneously exploited for Civilian and Defense utilization purposes.

An example of dual utilisation, showing a prominently increasing interest of Customers on a worldwide scale, is the Homeland Security. In this domain, the "typical" Defense needs of improving situation awareness and surveillance of national territory and borders, can also be shareable, up a given extent, with the homeland surveillance needs and security goals of the civilian Institutional and governmental organizations, such as police forces, magistrate, civilian protection, coast guards, etc. The "need to see" has recently accelerated due to the consciousness of Earth environment degradation, and of space observation as a privileged asset to maintain a strict surveillance on environmental changes, whose effects are interrelated on a world-wide scale, and to establish the limits of a sustainable development compatible with ecosystems fragile needs, as well as to monitor national territory and crisis areas to ensure citizens security and safety.

Hence, governments are currently investing in space-based systems for Earth Observation, Secure Communication, and Navigation Support in order to face these problems. This overlap of applications between Defense and civilian needs constitute the dual-use extent of the aforementioned space components, for a wide spread of Users as shown in Figure 9.1.

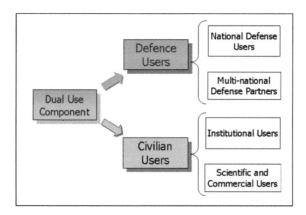

Figure 9.1 Dual-use system users

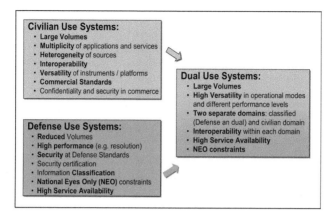

Civilian Use Systems:
- **Large Volumes**
- **Multiplicity** of applications and services
- **Heterogeneity** of sources
- **Interoperability**
- **Versatility** of instruments / platforms
- **Commercial Standards**
- Confidentiality and security in commerce

Dual Use Systems:
- **Large Volumes**
- **High Versatility** in operational modes and different performance levels
- **Two separate domains**: classified (Defense an dual) and civilian domain
- **Interoperability** within each domain
- **High Service Availability**
- **NEO constraints**

Defense Use Systems:
- **Reduced** Volumes
- **High performance** (e.g. resolution)
- **Security** at Defense Standards
- Security certification
- Information **Classification**
- **National Eyes Only (NEO)** constraints
- **High Service Availability**

Figure 9.2 Dual-use architectural principles for an Imagery Space System

The architectural principles of a dual-use space component can be conceived as a "synthesis" that compounds the characteristics of two domains, namely the civilian and the Defense one.

Figure 9.2 shows an example of this architectural synthesis applied to a dual-use space-based Imagery System.

The figure shows (in the light blue box) the needs related to the civilian use, which privileges the aspects of **multiplicity** of applications and services, in order to cover the different application purposes of heterogeneous typologies of civilian users (e.g. Institutional, scientific, commercial), hence the utilization of large data volumes, taken from heterogeneous data sources (e.g. images generated from a multiplicity of sensors, cartographic data from a Geographic Information System, etc.). In terms of architecture, the civilian domain requires **interoperability** among different entities, being them either information sources (e.g. an imagery space system) or an User utilization centre, in order to share the resources and to optimally provide global integrated services to the Customers. Finally, in the civilian domain, the systems are generally designed in order to guarantee confidentiality and security of the treated information according to commercial standards.

On the other side, the defence use systems (shown in the green box) normally handle data volumes smaller than an analogous civilian system, but the global service require higher system **performances** (e.g. provision of images at higher resolution), **reliability**, and **availability**. The system security aspects are also more "complicate", relying on Defense Standards and information classification level, and generally requiring a **certification** process to guarantee the compliance of the system to these standards.

The **synthesis** of the key architectural characteristics of Civilian and Defence domains defines a Dual Use system, which shall have an architecture that fulfils the most restrictive constraints of both domains. As such, a dual use system achieves large operational profiles such to allow large volume of imaging data takes on different targets, with an highly versatile architecture, capable to provide a variety of images, taken in different modalities, characterized by different sizes and performance levels such to fulfil different utilization needs of both domains. A dual-use architecture is generally partitioned into two domains: one exclusively dedicated to Defense use, and the other dedicated to civilian use, having few security-controlled interfacing points between the two's. High Service Availability also characterize of the dual-use components.

9.3 The Integration Concept in Dual Use Space Systems

The Space Technology supplies a set of integrated assets for security in different applications, such as Security and Environmental monitoring. Examples of space applications aimed at *Security* are the protection against threats of Strategic objectives such as national borders, shoreline, areas of interest; definition of plans for prevention and intervention in emergency situation, and simulation and training of security force or civil protection based on monitored and elaborated data (e.g. 3D images). The illegality prevention can also be based on advanced studies and simulation of the illegal traffics with location and classification of the threats (illegal traffic, illegal immigration, terrorism) through collection, interpretation, integration and dispatching of data from heterogeneous sources. On this application, the dual-use space assets provides formidable means for managing crisis, through continuous monitoring of the interest zones (theatre), identification of targets and objects, and support to intelligence activities.

The management of Environmental Risks can be conceived in the three distinct phases of: (1) strategic prevention, (2) "early warning", and emergency/crisis management, and (3) management of the post-crisis (assistance and reconstruction systematic activity), in catastrophic events which flood, fires, landslides, earthquakes. The dual-use space assets can support operations during each aforementioned phase. For example, the prevention phase can profitably use space resources for the necessary analysis of vulnerability, by checking points of failure inside the communication systems, transport infrastructures and distribution networks, in order to guarantee natural integrity against threats or catastrophic events.

The strategic objectives and applications, above indicated, are achieved through integrated systems that compound dual-use space systems and information sources (depicted in Figure 9.3) and support:

- **Earth Observation**: acquisition and elaboration of site images using heterogeneous sensors with panchromatic optical, multi-spectral, radar (SAR) and collection of meteorological information to support environmental monitoring. Further information and data (cartographic, climatic maps, data fusion) will be used as support to space observation.
- **Telecommunications** through integrated networks to satellite links, with mobile and fixed station and additional centres for multi-medial services towards civil and defence users.

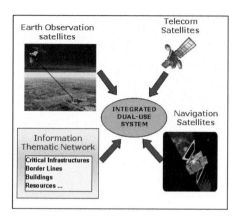

Figure 9.3 Integrated information services and space systems for dual use

- **Navigation**: concerning the development of an infrastructure to support navigation services for localization for different levels of accuracy for civil and defence customers.
- pre-existing information and data sources (e.g. GIS data bases, meteorological centre.).

From the above considerations, an integrated system for dual-use is actually founded onto a synergic combination of different space technologies, information systems, data fusion and processing techniques capable to achieve *a global system performance* provided to Users for different utilization purposes.

A pictorial view of an integrated system for dual use is shown in Figure 9.4, encompassing heterogeneous space systems, which are capable to interact among them and with terrestrial facilities and utilization centres managing the global provision of services to the end-Users.

9.4 Thales Alenia Space Italia Contributions to Space-based Dual Use Systems

Thales Alenia Space Italia is largely contributing, as World leading Company, to design, develop and put into operations Space Systems devoted to dual-use. As such, technological challenging systems are currently on our workbench or being commissioned for operations in the different domains of Earth Observation, Telecommunication, and Navigation.

Figure 9.5 represents the most significant system insofar built for **space-based dual-use Earth Observation** purpose: **COSMO-SkyMed**, shown in a pictorial view (figure 9.5a), and the constructive details of the active phased array antenna (figure 9.5b), and the satellite platform (figure 9.5c).

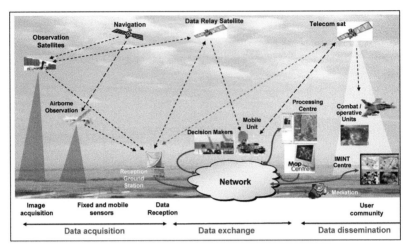

Figure 9.4 Example of integration of dual-use space and ground assets into a "system of systems"

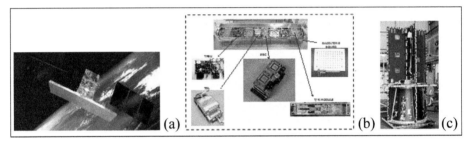

Figure 9.5 Dual Use Space Assets and Technologies for Earth Observation — **COSMO-SkyMed**

COSMO-SkyMed is the largest space system for radar observation, on a worldwide scale, which supports dual-use, through:

- A versatile multi-mode SAR instrument with wide variety and high volume of data images with different size, resolution, elaboration level, and accuracy (e.g. geo-localization) suitable for different application needs. The SAR antenna is the key technological component that implements the dual-use, since it support several imaging modes, thus fulfilling both Defense needs (i.e. very high resolution) and Civilian requirements (large coverage).
- The Payload Data Handling and Transmission (PDHT) is the other key on-board device that contributes to dual-use implementation through separate ciphering paths for Defense and Civilian/ Commercial data encryption, and allowing the multiplicity of use managing each image as a data file which can be separately handled and transmitted to its user's destination.
- The Satellite Platform providing agility, robustness, and resources necessary to sustain the highest operational profiles necessary to dual-use observation.

Figure 9.6 Dual use space assets and technologies for communications: **SICRAL** (figure a) and **ARTEMIS** (figures b, c)

- The System Security as a key enabling technology necessary to develop and qualify a system capable to provide imagery to Defense within the requested confidentiality and security constraints.

Figure 9.6 shows pictorial views of the most prominent realizations of Thales Alenia Space Italia for **space-based dual-use telecommunication**: SICRAL (figure 9.6a) and ARTEMIS (figures 9.6 b, c).

Dual-use telecommunication capabilities aim at ensuring services in regions during crisis or disaster occurrence, despite of possible failure of other ground communication systems, ensuring continuity of communication necessary for planning and coordination of operations through widespread data transmission capabilities (data, voice, video) provided to remote centres and operational forces deployed on field. Also, the "roadmap" toward dual-use telecommunications allow, while containing development costs, the integration of already existing assets with new generation systems. Thales Alenia Space Italia is fully responding to this challenge. The **SICRAL** (*"Sistema Italiano per Comunicazioni Riservate e Allarmi"*) is a cornerstone of space-based telecommunication for Defense and dual-use, that guarantees tactic and strategic mobile communications to several utilizations (land, naval or aerial), necessary to support crisis management, humanitarian relief, theatre operations, and peacekeeping missions. The aforementioned dual-use functionality is achieved through the new SICRAL 1b satellite, which provides capabilities of multi-load and multi-transmission in several frequency bands (e.g. EHF, UHF, SHF), counteracting against external disturbance or jamming, and high capacities such to satisfy the progressively increasing demands from Italian Defense, civilian institutions (Police Force, Civilian Protection, Fire department, Guardia Forestale, Cost Guard, Guardia di Finanza), NATO Command Network, and to some Alliance's Member States.

Another fundamental asset for space-based communication is **ARTEMIS** (Advanced Relay and Technology Mission Satellite), which is designated as a "Data Relay Satellite" (DRS), meaning that it acts as an interconnection node between the Earth receiving stations and a number of "client" satellites placed in Low Earth Orbit (LEO). It provides large coverage, and real time retransmission of data acquired by Remote Sensing Satellites to Ground Stations placed in Europe and Northern Africa. The above characteristics make ARTEMIS as a precursor of the development of future European space-based relay systems providing this services to "client" LEO satellites for both Defense and/or civilian missions. ARTEMIS has proven state-of-art technologies and advanced solutions, such as the S-Ka band Data Relay (SKDR), which is a radio frequency system used for data relay in Ka Band, and the Semiconductor Laser Inter-satellite Link Experiment (SILEX), which is the laser-based optical system used for inter-satellite data transmission, allowing higher data-rate, low power consumption, and reduced interference.

The following figure represents the most ambitious worldwide project for **space-based Navigation support**: the **GALILEO** system (Figure 9.7).

The GALILEO Program, managed by the European Community in the framework of the Direction of Transports and Energy (DGTREN), and technically supported by ESA, aims at launching a constellation of thirty satellites placed in Mid-Earth Orbiting (MEO) at approximately 23000 km of altitude, and able to provide a worldwide coverage to the users for on-ground as well as aerial and marine navigation. Although GALILEO is not natively conceived as a Defense Program, its design that takes into account high security measures, which enable it to operate as dual-use system which control is given to a Civilian Authority. Indeed, GALILEO is capable to provide its services with different level of accuracy, such as: (1) the Open Service (OS) will provides positioning, navigation and timing signals that can be accessed free of direct charge, suitable for mass-market navigation applications (e.g. in-car navigation, positioning with mobile telephones); and (2) Commercial Service (CS) through encrypted signals which will guarantee a sub-metric accuracy. These GALILEO services will be of primary importance to solve the security problems related to the mobility and transportation, e.g. the monitoring and control of the aerial traffic as well as the support to the mobility, with services provided at a guaranteed Quality of Service (QoS), in terms of data integrity and reduced response time necessary to support Defence applications. Also, it's worth to mention the GALILEO support to the cellular network by means of localization services to the user's mobile unit, which will allow to provide value-added information (e.g. the localization of a hospital or other points of interest in the vicinities) in support to humanitarian operations.

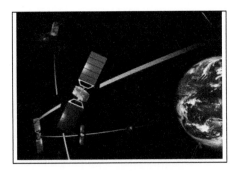

Figure 9.7 Dual use space assets and technologies for navigation — **the GALILEO system**

9.5 Conclusions

The rising interest of Customers toward space-based dual-use systems is strictly related to the progressively increasing demand of security from government and institutions, together with the actual need of containing and sharing the system development costs among different subjects. Thales Alenia Space Italia has promptly answered to this planetary challenge by promoting and developing design solutions and technology advances for dual use space systems. This is also due to the remarkable impulse, first in the world, that the Italian institutions, such as the Ministry of Defense and the Italian Space Agency, are giving to the development of dual-use technologies.

The results obtained from these efforts are highly encouraging: nowadays dual-use is a reality in the most important applicative fields of surveillance from space, environmental monitoring, navigation support, and secure telecommunications.

Another fundamental aspect, mentioned in this chapter, is the integration of heterogeneous space assets into a larger context of an integrated geospatial "system of systems", capable to provide the End User Communities (Defense and civilian) with capabilities at insuperable global performance. Indeed, all the space systems delineated in this chapter are actually providing interoperability and expandability features in order to ease such an integration.

These dual-use space systems characterise the best Italian contribution to the enterprise of augmenting homeland security, preserving the environment, and ensuring the safety of the citizens, through systems and technologies that make feasible today the challenges of the future.

Chapter 10

Electro-Optic Sensors for Dual Use Space and Airborne Applications

Renzo Meschini, Armando Buccheri

Galileo Avionica

10.1 Introduction

In the present situation, characterized by difficulty in allocating in the national budgets funds for the development of advanced technologies, the role of the Aerospace Industry is becoming even more important. In fact, it is unquestionable that many technologies and processes, developed along the years for challenging space missions or military programs, have generated industrial patents for the production of consumer goods. Nevertheless, even this model that uses the technology transfer from "high class" application (space and military) to "low class" application (industrial and consumer), has to be improved. This is due also to the fact that the cost of the aerospace products, that until some years ago was not so determining, today has assumed the same importance as for all other industrial products.

Then the challenge for the Industry is to identify the emerging technologies and to invest on them, trying to exploit from the very beginning all the potential applications, with the purpose to develop products even more performing, reliable, and cheaper, suitable for both military and civil contexts with minimum adaptations.

As evidence of this new approach, on which Galileo Avionica firmly believes, we describe in the following some examples of dual technologies developed in our Company.

10.2 Electro-Optical Sensors

Many Equipment and systems produced by Galileo Avionica have an intrinsic dual use nature. Key technologies for radar, avionics, electro-optical, and UAV are employed for military, security and civil applications within products having very similar configurations.

Also Fully Military equipments are often used within civil environment: flight Control and mission management, computers for helicopters and UAV, surveillance radars, e-o sensing equipment, tactical UAV may be employed in search and rescue missions, civil protection operations and homeland protection environment.

Galileo Avionica acts in the field of avionics and electro-optical systems from many decades. It is specialised in providing the "eyes" and "the brain" for defence and security applications. These products, that in the past had mostly military applications were mainly based on opto-mechanics. Today they have largely widened their functionalities thanks to the application of Information Technology (IT).

IT actually has a crucial role to improve the performance, reduce size and mass, increase the capability of analyzing, processing and storing. This evolution of electro-optical technology is an example where the progress of the military industry, has been favoured by investments made in the civil one.

Another factor that contributes to reduce the distance between military and civil applications is the increasing request of performance emerging from the recent attention to the Security problems that need the deployment of a large amount of equipment and sensors for the use by civil and paramilitary Authorities.

An example in this sense, derived from the experience of Galileo Avionica, is given by the EOST system, that, integrating different optical systems, is employed both in military missions (navigation, targeting) and in patrolling and surveillance missions. The EOST system is also suitable for applications in the field of the environment monitoring as well as in the field of scientific research, representing a typical dual use system.

Another example where technologies developed for military purposes have been transferred into a product for space application is represented by Laser Transmitters. In this case the investments made in the development of rangefinders have created the know-how necessary to successfully compete in the development of challenging transmitters based on Nd:YAG lasers that are adopted by ESA in two space instruments: ALADIN and ATLID. The former is a Doppler wind lidar able to measure the velocity field in the stratosphere and troposphere from satellite, the latter is an atmospheric lidar for the accurate measurement of the atmosphere composition.

Figure 10.1 EOST 45

Figure 10.2 ALADIN Transmitter

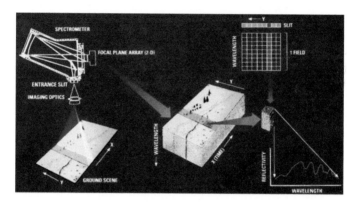

Figure 10.3 Principle scheme of hyperspectral

10.3 Hyperspectral Technology

Hyperspectral Imagers (HSI) use spectrometers coupled with 2-D detectors to produce a set of images of the same scene in a continuous (typically hundredths) of spectral bands. The spectral dimension and the spatial dimension across the track are provided by the detector, while the spatial dimension along the track is provided by the motion of the platform (pushbroom concept). It does not require moving parts to scan the scene. The output of a HSI is not a simple image but is generally referred as "*image cube*" containing both spatial and spectral information. In this way it is possible to measure the spectrum corresponding to each pixel of the scene. Figure 10.3 illustrates the principle of operation.

HSI allows to determine the spectral signature of the objects present in the scene:

Objects of the same shape but made of different materials can be distinguished, through their spectral signature, while they will appear identical in conventional images.

This capability allows an amount of application related to defence, security, civil protection, etc, for example:

- Detection of hidden and/or camouflaged targets
- Detection of underwater hazards
- Bathimetry of coastal waters
- Detection of dangerous contaminants and chemical agents in water and land
- Monitoring of land and sea resources
- Monitoring of natural hazards (fires, floods, slides)
- Preparation of thematic maps

The processing depends on the level of information that is required. Simple survey of the scene to detect spectral signature anomalies, that could be targets of interest, requires light processing and can be done in real time. Recognition of objects is done through the comparison of detected spectral signature with databases. It generally needs atmospheric correction of the images through the analysis of specific spectral bands and the use of radiative transfer models (i.e. Modtran). More quantitative analyses, like the estimate of abundance of minerals or pollutants, require specific algorithms.

Hyperspectral technology constitutes the best example of a development where all the possible applications (military, security, environment, science) have been taken into account. Galileo Avionica invests constantly in the development and in the continuous improvement of this technique from the '80 s. Initially this technology have been developed for space, mainly for scientific purposes. Examples are the instruments VIMS and VIRTIS, flown onboard four satellites (Cassini, Rosetta and Venus Express — ESA and Dawn — NASA) to provide spectral images useful to study the atmosphere and the soil composition of planets, asteroids and cometary bodies.

Hyperspectral technology has been also exploited for the development of GOME, an instrument dedicated to the measure of gases present in the earth atmosphere, with special attention to ozone. The instrument, flown in a first version in 1995 on board the ESA satellite ERS-2, is today at its second generation flying from 2006 on board METOP-1. GOME measures the ultraviolet and visible radiation scattered by the atmosphere, and through its spectral analysis is able to recognize the different gas molecules and to quantify their concentration. It is specially able in measuring the stratospheric ozone layer that protects against the dangerous UV radiation and that contribute to maintain the equilibrium in the radiative processes that regulate the earth temperature.

Figure 10.4 VIMS Visible InfraRed Mapping Spectrometer

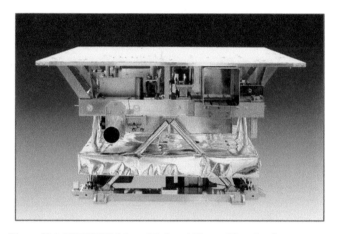

Figure 10.5 VIRTIS Visible and Infrared Thermal Imaging Spectrometer

At present this background of know-how, technology and capabilities has been transferred to the development of a new instrument: SIM.GA — Sistema Iperspettrale Multisensoriale Galileo Avionica. It is an avionic hyperspectral system equipped with two optical heads operating in the VNIR (0.4–1.0 micron) and SWIR (1.0–2.5 micron). The complete system includes also an IR camera operating in the range 8–12 micron. The system is equipped with an electronics system for control, data acquisition and data processing.

The variety of applications of this new sensor makes it a typical example of dual technology: from environment monitoring to the military intelligence, from thematic mapping to target detection in military or security missions, from observation of crisis theatres to post operation evaluation. The applications are virtually infinite and depend mainly on the ability of potential customers and users to collaborate with our Company in the development of dedicated databases and processing algorithms specific for the selected application. For its part, Galileo Avionica is

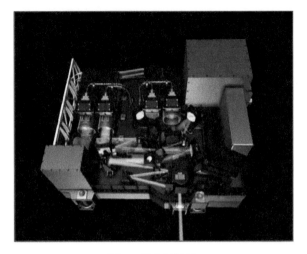

Figure 10.6 GOME

Figure 10.7 Electronics and Optical Head

aiming at the maximum flexibility of the system, not only for what concerns the performances and the hardware configuration, but also with respect to the capability of integration on a ample variety of platforms: SIM.GA in fact can be accommodated on several aircrafts, helicopters and UAV.

Several flights of SIMGA, the GA avionic hyperspectral sensor prototype, have been already performed and are planned in the near future, both for characterizing the physical and chemical characteristics of natural targets, or to identify "anomalies" in the observed scene: the dual use is trivial in this case. The goal in the GA strategy is to propose to the dual market an "instrument system" based on core hardware, the same for military and civil application, with specialised functions, mainly realised via software, able to satisfy a range of various needs. The military applications that are currently tested include Intelligence, Surveillance, Target Acquisition and Targeting.

Figure 10.8 SIM.GA onboard P66

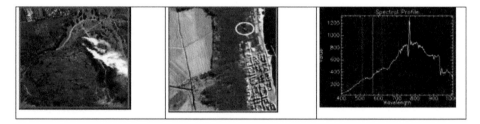

Figure 10.9 Examples of hyperspectral images

At the moment SIM.GA has flown a first time onboard a CASA 212 aircraft for testing the system with the acquisition of images and data along the Tuscany littoral. A following series of flights has been funded by the European Space Agency and in cooperation with INGV, for the acquisition of data in support to the development of GMES; particular interesting results have been found in the monitoring of volcanic areas. In that occasion SIM.GA has been mounted on a small aircraft. At present SIM.GA is supporting the development of flight campaigns, made by TELAER on a P66, for the Italian Ministry of Agriculture. Figure 10.8 shows the accommodation of SIM.GA on the plane.

Figure 10.9 shows a typical output of the instrument. On the left an RGB image which includes a fire, at the centre a false colour SWIR image and at the right the spectrum corresponding to the encircled zone. The spectrum reveals the presence of potassium associated with the burning of vegetal material.

10.4 Hyperspectral for Space

A version for earth observation from space is also under development in the frame of the ASI program PRISMA. PRISMA is a demonstrative mission aimed at the environmental monitoring, but it offers also the opportunity to test on the field military applications that could be exploited in dedicated mission as, for example, the next generation observation satellites MUSIS.

In conclusion, hyperspectral technique represents a good example of dual use of an electro-optical sensor, both from aircraft and from space. For aircraft a prototype is already available and GA is available to cooperate with Defence to evaluate its application. For space, the ASI demonstration mission PRISMA will offer the possibility in a short time frame to put available to any kind of Users hyperspectral products to be tested and evaluated in the perspective of a following operational service.

In particular it is our firm intention to propose a dual use mission as PRISMA follow up, constituted by a small constellation of satellites, on the same pattern as COSMO-SkyMed.

Chapter 11
TETRA in Heterogeneous Networks

Marco Tommasi, Giovanni Guidotti, Nunzio Cocco

SELEX Communications

11.1 Introduction

The recent events occurred in the first part of the new century have shown how the threat of Global Terrorism Attack and the continuous increase of emergency situations, such as peace-keeping missions or civil protection missions after natural disasters, have brought the military and the civil worlds closer. In fact they have demonstrated that nowadays an effective **Crisis Management** can be granted only through the exploitations of both military and civil technologies and applications into **Dual Mode Systems**.

In order to achieve this goal a key topic to be addressed is the advanced implementation of the so called **"Integration concept"** (I-concept). The I-concept means that human and electronic entities can exchange all types of information through different communication networks, also allowing different applications to perform data fusion and extract important operational indications.

In the development of a communications system the I-concept would bring to the conceivement of a **Heterogeneous Network** through the effective exploitation of wireless and wired connections, satellite and terrestrial technologies, to provide the optimal performance.

In this frame, **TETRA** (TErrestrial Trunked Radio) networks and **Satellite**-based systems are two necessary building blocks of an I-concept-based network devoted to both civil and military scenarios. The synergic combination of the two complementary technologies allows, on one hand, the quick deployment of mobile networks providing flexible, yet robust and secure communication by means of the TETRA cellular nodes, and on the other hand, the long range reliable and secure connectivity in the absence of any terrestrial infrastructures through the use of satellite systems.

11.2 Crisis Management — A Complex Process

Crisis management is a complex process in which various entities have to interoperate (Police Forces, Civil Organizations, Military Forces) and different communications means and technologies need to converge:

The integration of Neworks and the interoperability among various communications systems could be beneficial in many situations, but certainly becomes crucial

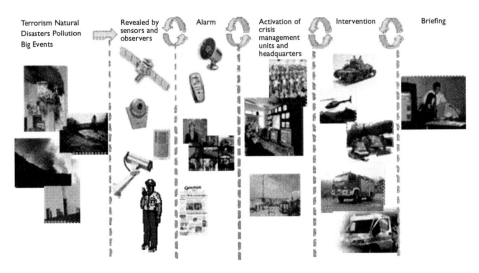

Figure 11.1 Integration of Networks, a key topic to face Crisis Management

when facing an emergency. This is a typical dual mode application, including both military and civil environments.

11.3 The Integration Concept

An I-Net can be depicted through the layered architecture provided in Figure 11.2. The global model envisages **a terrestrial layer** made by network nodes and connections and an **aero-space layer** based on satellite systems and aerial components which can be usefully considered to complement coverage, provide additional services and a back-up to the terrestrial section.

The aero-space component is itself structured in various layers: the HAP (high altitude platform) layer, where manned or unmanned stratospheric vehicles are located at about 20 km height, and the layers where satellites can be mainly located, low, medium and geostationary orbit (LEO, MEO and GEO).

The Satellite-based systems are a fundamental building block when we have to provide real time connectivity into emergency areas where communication terrestrial infrastructures are not available or have been damaged by natural or intentional disasters.

The Terrestrial Layer should be based on the integration of wireless and wired connections and the interoperability among various communications means. In this frame, TETRA (TErrestrial Trunked Radio), the first truly open standard for the digital mobile radio system, allows the quick deployment of mobile networks: speed and mobility are in fact the two key-words particularly in an emergency environment. Furthermore, the interesting feature of mobile-to-mobile connections is a plus of the TETRA technology, based on a very powerful operation mode, the

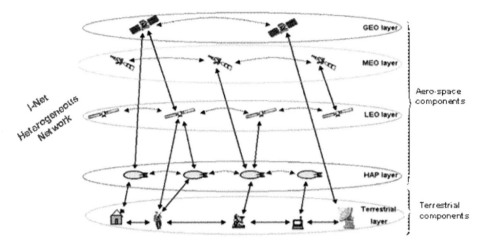

Figure 11.2 The I-Net architecture: terrestrial and aero-space components

DMO (Direct Mode Operation), that allows to deploy an ad-hoc network without the need of any infrastructure.

11.4 Heterogeneous Network

A first step towards Heterogeneous Networks interoperability can be achieved by convergence over an IP core network: all voice, data and video signals can be exchanged through different networks using an IP core.

Today, an intense work is ongoing to integrate TETRA networks with different technologies characterized by different maturity levels, also in the perspective of meeting the user growing demand for a wide range of multimedia applications.

TETRA network is then integrated with other legacy wireless communications means (HF, VHF, Simulcast) and with new high speed wireless networks (WiFi, WiMAX) and public networks (GSM, UMTS). All networks interoperate through an IP core, where the various interworking functions guarantee a set of communications services. Additionally, long distances as for operation in remote areas can be covered by using satellite connections. (Figure 11.3).

11.5 Tetra Today: Status and Objectives

TETRA is the first truly open digital private mobile radio standard. TETRA is defined by the European Telecommunications Standards Institute (ETSI) that joins the forces of network operators, national administrators, equipment manufacturers and users. TETRA standard does not specify a detailed network as in case of other standards; its aim is to define the air interface and the interface between the TETRA network and other networks like ISDN, PSTN, PDN, PABX and other

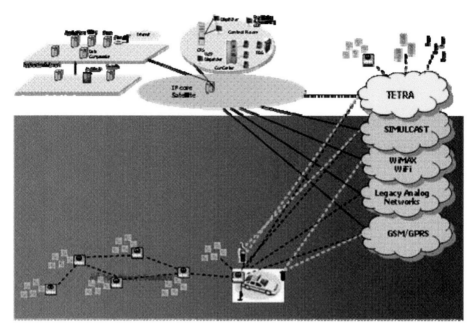

Figure 11.3 Heterogeneous Networks architecture

TETRA systems. The network architecture is left to proprietary implementations in order to provide optimized solutions for various applications. TETRA is, hence, a powerful multi-function mobile radio standard that provides a comprehensive tool kit from which system planners may choose in order to satisfy their requirements. The TETRA suite of mobile radio specifications provides a radio capability encompassing trunked, non-trunked and direct mobile-to-mobile communications with a range of facilities including voice, circuit mode data, short data messages and packet mode data services. In addition to these basic services, it supports a wide range of supplementary services such as call authorization by dispatcher, dynamic group assignment, priority calling and late entry to calls in progress. A discreet listening facility allows the control room to monitor calls. If a vehicle is stolen or hijacked, its TETRA radio can be remotely switched into ambient-listening mode to monitor verbal exchanges. Many of these services are exclusive to TETRA.

TETRA has now been adopted in many countries as the de-facto standard for new digital trunked radio systems and there has been a rapid world-wide growth in the implementation of networks in both the public and private sectors. Therefore, it is very interesting to consider the current positioning of TETRA in the marketplace, since the technology is capable of providing both specialized high grade services for the most demanding users, such as the military, and it is also able to compete in the general business sector as a cellular service. The versatility of TETRA is evident in its interest across several market sectors, such as Public Safety, Public Access, Transport, Utilities, Cellular service provision, Military.

Within Europe, many public safety TETRA networks are being commissioned, mainly because of the interoperability requirement. In addition, there is a growing interest from the military sector where users are keen to deploy commercial equipment to introduce state-of-the-art technology deriving the benefits of economies of scale. The growing interest in TETRA technology is demonstrated by the noticeable number of large national systems being deployed worldwide, and by its increasing use in various application areas.

11.6 Integration of Tetra and Satellite Systems

The benefit of the integration between TETRA and satellites is mutual, as depicted in Figure 11.4: the satellite-based system(s) can provide to the TETRA-covered area data/information that can help TETRA users. The data/information exchanged among TETRA users could be rendered available to satellite-covered areas.

This approach is the best solution to solve communications needs in presence of Crisis Management, such as:

- peace-keeping mission, when an armed force is moving into an hostile site where the presence of a network infrastructure is not guaranteed;
- natural disasters, such as fires, storms, Earthquakes etc., where a team need to keep in touch each other without needing any external facility.

TETRA could be connected in principle with various satellite systems, characterized by different services and/or coverage and/or nature (civil or military).

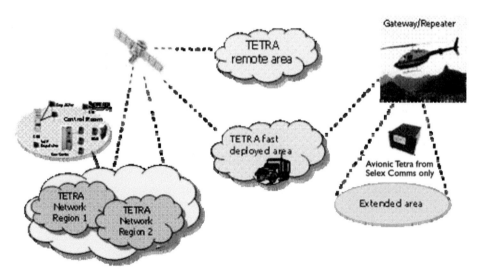

Figure 11.4 Example of Satellite comms and TETRA integration

In particular, data can belong to:

- Global Navigation Satellite Systems (GNSS), i.e. GPS, EGNOS, GALILEO in the near future;
- Global or Local Communications Satellite Systems (GCSS, LCSS) in either broadcasting or point-to-point/multipoint connection modes;
- Global Earth Observation Systems (GMES and — in the medium term GEOSS);
- Global Integrated Satellite Systems (GISS) where two or all three types of data (communications, navigation, Earth observation) are available (medium and long term vision).

In the above frame, TETRA could be a flexible and quick means to deploy a local terrestrial "connection area" that, thanks to a proper interface (TSI, TETRA Satellite Interface) and, hence, to the satellite connections, can more easily perform its task in disaster areas where the terrestrial networks have been damaged.

11.7 Conclusions

The integrated platform described above is the best solution for managing dual use scenarios, like peace-keeping missions and emergencies. It provides a timely Early Warning and an effective Crisis Management, based on the following:

- a synergic combination of TETRA and Satellite Networks, as the two fundamental building blocks;
- the integration with other communications systems, such as HF, VHF, Simulcast and with new high speed wireless networks (WiFi, WiMAX) and public networks (GSM, UMTS), to form an HETEROGENEOUS NETWORK.

Key benefits of employment of TETRA into Heterogeneous Networks:

- Long range satellite connectivity combined with high mobility TETRA Networks;
- Rapid deployment of communications media in emergency conditions;
- Enhanced secure communications provided by TETRA cellular nodes and satellite secure links;
- Integration and interoperability with Heterogeneous Networks focused on specific applications, to form an integrated and effective "early-threat detection network".

Chapter 12

Trends of Enabling Technologies for Sensors, Systems and System of Systems

Marina Grossi

SELEX Sistemi Integrati

12.1 Introduction

This chapter will focus on trends in emerging technologies. Currently and in the future, these technologies will enable continuous innovation and revolution in the sectors of sensors, systems and systems of systems. Through its R&D efforts, SELEX Sistemi Integrati (or SELEX SI) is a leader in the sector and is continuously engaged in supporting and developing its state-of-the-art products and systems.

After a brief history of SELEX SI (Par. 12.2), the chapter highlights the transformation of sensors, systems and systems of systems as part of an ongoing evolutionary process (Par. 12.3). This is followed by short descriptions of the company's core assets such as active antennas (Par. 12.4) basic technologies, multi-core processing (Par. 12.5), software and middleware (Par. 12.6). Moreover, as part of a revolution process, this chapter also outlines emerging "killer technologies" such as photonics and nanotechnologies (Par. 12.7) which require multidisciplinary approaches, high-risk investments and "creative destruction" of established technologies. These "enabling technologies" will lead to new solutions and improvements of several orders of magnitude. They *"could open a new era by enabling, supporting and driving the 21st century knowledge-based society"*.

12.2 SELEX Sistemi Integrati

12.2.1 SELEX SI at a Glance

SELEX SI, a Finmeccanica company, designs and develops large systems for homeland protection, systems and radars for air defence, battlefield management, naval defence, air and airport traffic management, coastal and maritime surveillance.

With a fifty-year track record in system integration, SELEX SI has a customer base in no less than 150 countries and complete mastery of the relevant advanced technologies — the company's true wealth and expertise — as well as an extensive commercial offer of state-of-the-art solutions, backed up by full logistic support.

With over 3,000 employees and sites in Italy (Rome, Fusaro and Giugliano in the Naples area, Genoa, Taranto and Pisa), the UK (Portsmouth), Germany

(Neuss-Rosellen) and the USA (Kansas City), SELEX SI is a leader in research and development thanks to annual investments of up to 20% of the production value.

SELEX SI is a systems engineer preferred supplier for air defence and battle-field management; a leading player in the field of Large Systems for Homeland Protection; designer and developer of the world's most complete coastal surveillance Vessel Traffic System; Europe's leading producer of 3D NATO-class long-range radars; integrated supplier of the "Airport Global System"; and a leading player in naval defence systems.

12.2.2 SELEX SI Heritage

In 1951, Finmeccanica and the US company Raytheon created Microlambda, to operate in the field of applied electronics and producing radar systems under license for land and naval applications. In 1956, a new company, SINDEL, was set up with capital provided by Edison; several engineers and technicians from Microlambda moved to the company which was specialised in professional electronics.

In 1960, an international agreement led to the creation of SELENIA — Associated Electronics Industries — whose shareholders were: 40% Finmeccanica, 40% Raytheon and 20% Edison. Over the years, the name "Selenia" would gain worldwide recognition for civil and military surveillance systems.

In 1990, Selenia merged with Aeritalia, a leading company in the aerospace sector. The merger gave rise to ALENIA, with more than 30,000 employees, dealing with aeronautics, radar, naval systems, missiles, space applications, aircraft engines, as well as systems for environmental protection. Afterwards, the Alenia Difesa management decided to set up an international alliance to deal more effectively with the fierce competition in the defence electronics sector and to expand into other more demanding markets.

The chosen partner was the British firm GEC-Marconi, and ALENIA MARCONI SYSTEMS — AMS came into being in 1998 with shareholding divided equally between Finmeccanica and GEC-Marconi. In 2005, the joint venture agreements expired and Finmeccanica acquired 100% of the Italian shares of AMS.

The new firm, SELEX Sistemi Integrati, included activities from BAE Systems in Air Traffic Management and Air Traffic Control in the United Kingdom, from Gematronik Germany in meteorological radars, and from ASI in airport navigation aids in the USA.

12.2.3 SELEX SI Global Systems and Sensors

With extensive experience in the different areas of the industrial defence sector, and applying its own skills and know-how to large system architectures, SELEX SI is positioned as a global supplier of defence systems, from C4ISTAR systems to ground radar sensors and from naval combat systems to onboard ship sensors.

SELEX SI's offer can be broken down into three complementary and synergistic business lines: air defence and battle space management systems, naval defence systems, and integrated customer support.

SELEX SI has a specific unit devoted to customer support activities, from traditional after-sale to integrated support, so as to ensure assistance in maintaining and upgrading customers' systems and prompt response to any requirement. The main services include integrated after-sale support, after-warranty logistic service, tailored after-sale support and training courses.

To respond quickly to customer needs, SELEX SI has developed a customer-support web desk, with operators answering any kind of question and providing on-line solutions. We establish a partnership with our customers based on information exchange, shared management of projects and processes, in order to foster enhancement from today's systems to next-generation capabilities.

12.3 Evolution

12.3.1 Evolution of Radars and Systems

A century ago, Christian Hulsmeyer took out a UK patent on the first idea containing the "DNA" of the sensor, now generally known as RADAR (RAdio Detection And Ranging). Then, it took more than thirty years to put the idea into practice. The main achievements in radar performance were made by Guglielmo Marconi in 1937, and later by Prof. Ugo Tiberio with the GUFO radar.

In this continuous evolution, technologies related to antenna and signal processing have increasingly absorbed all the other ones like bulk transmitters, MF and RF. We envisage future digital radar with thousands of heterogeneous nodes, highly adaptive and cooperating in real time, with transmitted power generation elements and digital receivers contained directly in the antenna. Key elements are system and software architecture and communication, with software-based signal processing. These should provide very demanding algorithms such as adaptive hard real-time digital beam forming, or knowledge-based information fusion, among other heterogeneous sensors.

New radar sensors could be multifunctional such as the EMPAR (European Multifunctional Phased Array Radar) capable of performing search, surveillance and tracking operative modes, with hard real-time adaptivity. In addition, we are developing "multi-domain" sensors where the radar operative functions, the passive and active electronic warfare functions, and the communications functions are managed and completely fused in the same system.

12.3.2 Evolution of Functions and Algorithms

Techniques, technologies, systems and applications of radar are extensively described in many papers, books and conference proceedings. For instance IEEE Trans. on Aerospace and Electronic Systems and IEE Proc. Radar, Sonar and Navigation are some of the key reference journals.

The key feature in modern radar is a complete and hard real-time adaptivity to the environment; active and flexible electronic scanning antennas, adaptive signal and data processing based on software and middleware ensure increasing radar quality.

In the radar system architecture and algorithm arena, a more recent major advance is the application of knowledge-based expert systems (KBS). KBS is in the realm of artificial intelligence and consists of a knowledge base containing information which is specific to a problem domain, and an inference engine that employs reasoning to yield decisions. KBS systems have been built: some are very complex with thousands of rules while others, relatively simple, are designed to tackle very specialised tasks. These knowledge and expert rules may be used to select the radar operation mode, the algorithms and training data, thereby significantly improving the performance of modern adaptive array radar in dynamic and non-homogeneous environments. Sources of data are: digital terrain model, surface cover, geographic maps, meteorological condition, data from other sensors, etc. It is expected that knowledge about local radar environment will enhance the ability of radar to maintain multiple tracks through areas of shadowing, clutter and other sources of interference that might be present in the surveillance scene.

For example, KBS adaptive signal processing can:

- address the actual environment and perform better than conventional algorithms that hypothesize IID conditions (Identical Independently Distributed) for clutter data
- ensure that shadowed range cells are excluded from training data
- determine the choice of data to best match the interference scene using information from diverse sources
- improve many different functions such as Space Time Adaptive Processing STAP and Rotational Motion Compensation functions.

12.3.3 From the ATCR to the Airport Global System

In the Civil Systems arena, our innovative solutions have to ensure efficient and safe management of airspace and airport traffic. SELEX SI is an integrated supplier of the "Airport Global System": from weather radar systems to ATC radars, from airport-ground manoeuvres to smooth air traffic management, the company offers a complete product and service portfolio by putting itself forward as the ideal technological partner for turnkey solutions.

SELEX SI has an established customer base in over 150 countries providing total Air Traffic Management capability from En-Route/Approach to Surface Management systems, Gate-to-Gate Surveillance with their family of ATC Primary and Secondary radars, new generation CNS/ATM systems such as Satellite Navigation Augmentation Systems, Automatic Dependent Surveillance and Controller Pilot Data Link Communications.

The company also provides navigation and landing aids by SELEX SI Inc. (ASI) and weather radar systems by SELEX SI GmbH (Gematronick).

To cope with the forecast growth in air traffic, the very philosophy of control is changing with the introduction of new communications, navigation and surveillance techniques. Active control, flight by flight and sector by sector, will give way to enhanced air traffic management.

CNS/ATM is "the adaptation of modern technology to enhance communication links between aircraft and air traffic controllers, improve pilot's ability to navigate the aircraft and to increase air traffic controllers' ability to monitor and control flights".

The introduction of CNS/ATM technology is a major change in operational requirements of Air Traffic Management and SELEX SI can manage this transition. We are also a full participant in the European Union, ECAC and Eurocontrol research programmes, helping define the systems and standards of the future.

12.3.4 Systems of Systems for Homeland Protection

Within the Finmeccanica Group, SELEX SI is the design authority and prime contractor for designing, developing and marketing systems of systems for Homeland Protection, a domain that includes both Homeland Security and Homeland Defence solutions.

This role endows SELEX SI with a wide portfolio of technologies, products and skills, both its own and of other Finmeccanica companies.

A system of system is a group of operators and systems capable of gathering and distributing information, enabling collaboration between the components, sharing the evaluation of situations, and automatically synchronising actions. A system of system has greater capacity and capability than the sum of individual parts, and its effectiveness increases as a result of coordinating and rationalising resources already present, while introducing new enabling technologies.

12.3.5 Technology Governance

Radars, sensors and systems are continuously evolving: what are the technologies required?

Until recently, technology-based companies prepared a key document in the business scenario: the Technology Plan, listing the technologies to be developed in order to allow the company to pursue its mission.

"Technology Governance" in SELEX SI (Figure 12.1), and in all Companies embracing the modern vision of integrated processes, may now be considered a wider approach to technologies. In fact, the concept of Technology Plan is part of a scenario flowing down from the Company Mission and Business Strategy to the Product Portfolio, the Product Planning and the Budget/Plan funds, and then to Technology Plan. This document is now named "Innovation and R&D Strategy Plan" in order to highlight the following concepts:

- the key aspect of innovation amongst the company's primary aims
- the strategic approach to technologies and their strategic value
- the positioning of the plan definition in a yearly-based virtuous process including it and the company's other major plans.

Figure 12.1 SELEX SI Technology Governance Strategy

We would like to highlight some aspects of the "Innovation and R&D Strategy Plan":

- the technology roadmap based of the product roadmap
- identifying the technology gaps to be covered in order to meet future positioning, starting from the current status relevant both to the Competitive Impact and the Competitive Positioning
- identifying competitors, opportunities and partners for technology development
- the features of the elements that will enable the company to be the prime mover for the integration of platforms within the Finmeccanica Group.

At the end, we have to remember that SELEX SI's Innovation and R&D Strategy Plan is part of Finmeccanica's "Technology Governance". It is therefore integrated with the equivalent plans coming from other Group companies. Until recently, technology-based companies prepared a key document in the business scenario: the Technology Plan, listing the technologies to be developed in order to allow the company to pursue its mission.

12.4 Active Antenna Technologies

12.4.1 Active Phased Array Radar

Phased array radar has a long history in Selex SI. A "passive" system was first conceived in the 1970s and is in production today.

As the technology has reached maturity level, complex systems based on Active Electronically Scanned Arrays (AESA) are the best solution for large ground, naval and avionic applications. AESA systems give complete control of antenna pattern, and the increasing use of digital beam-forming provides increased capabilities for multi-task adaptivity to operational conditions.

Active array architecture requires thousands of T/R modules positioned behind the array radiating elements. Typically, the cost of each TRM (from 100s to 1000s of euros) plays a significant role in determining the overall AESA radar cost.

The radar trend evolves through multiple and conflicting requirements driven by complex defence scenarios that include homeland security and multifunctional antenna systems. A technology that improves capability and flexibility at lower cost and weight will be the primary challenge. A view on the future is the development of advanced multifunctional systems which use a single antenna aperture to support multiple services across ultra-wideband frequency range and common signal processing. Here the introduction of new technologies such as GaN, photonics, compact array multilayer structure, and SiGe for multi-chip devices which also include digital functions, will offer significant advantages in achieving the required levels of performance and cost.

12.4.2 Transmit/Receive Modules

To be competitive against conventional radars, AESA systems need to fulfil the ongoing demand for increased performance, and above all must be competitive in terms of reliability and cost. For this reason, dominance in innovative Transmit/Receive Module (TRM) technology, the key enabling-technology in AESA systems, is of paramount importance and as such many defence companies are developing independent capabilities to ensure a competitive edge in this field. Such capabilities are critically dependent not only on the availability of a technology which enables the design and fabrication of highly compact, efficient and reliable T/R modules, but more important on well defined technology roadmaps and related "re-use core-module" concepts which ensure overall competitiveness in product life cycle costs.

A generic block diagram and a prototype example of a related X-Band "core-module" TRM are illustrated in Figure 12.2. These modules comprise: a T_x-chain for the generation of transmit power, a low-noise R_x-chain for signals received from the respective radiating element, a vector modulator for signal phase-shifting in T_x and R_x modes for beam steering and variable gain setting for aperture weighting during reception, and finally a duplexer for signal switching from T_x to R_x mode.

The design concept of such technology is based on a highly integrated solution utilising a minimum number of MMICs and robust "design-centring" to specifications via the introduction of appropriate control/calibration circuits and tuning-free integrating procedures.

To give TRM products a competitive edge, current and future development activity is based on a detailed "Technology Roadmap" — guidelines which extend from short-term requirements for current multi-function radars to long-term

Figure 12.2 Typical TRM block diagram and related example of X-band prototype

requirements for next generation wideband multi-domain/multi-role applications. On a short-term basis, the roadmap envisages: a reduced number of GaAs components via the introduction of multi-function MMICs, a high level of integration via LTCC/HTCC substrates for interconnect and packaging and the integration of digital functions, improved performance GaAs PHEMT power components, and design methodologies for simple low-cost "re-use core-module" implementation. On a long-term basis, the roadmap envisages: single chip (GaAs or GaN) RF Front-End and single chip SiGe Back-End solutions for high power TRMs, complete single chip SiGe TRM function for low power applications, innovative high thermal conductivity packaging materials for compact high power GaN components, and RF MEMS/MMIC integration for compact wideband modules.

12.4.3 GaAs and GaN Enabling Technologies

The strategic importance of T/R modules (which are often not available or subject to export restrictions) determines the need to directly control the key/critical enabling technologies. In recent years, the procurement of GaAs COTS components for professional electronic application, especially for the European Defence Industry, is proving to be less attractive, and in some cases at high risk. For many applications, the critical high performance components, such as GaAs HPAs, are often subject to export license restrictions, and even for the less critical ASIC components the need to schedule access to commercial foundries (military less than 4% commercial business) often increases development time and cost. For many years, against this background, the mission of the SELEX SI GaAs/GaN foundry has been to mitigate the limitations associated with the procurement of GaAs components for TR module development and production. As such, the foundry's activities has been devoted solely to the development and production of strategic MMICs, and in particular power HPAs and E/D mode multi-function vector modulator MMICs for space and defence applications. Currently, state-of-the-art C and X-Band Power PHEMT MMICs, based on 0.5 and 0.25 μm gatelength technology, are available. (see Figure 12.3).

GaN-HEMT technology is currently under development (see Figure 12.4) with the same objective for future applications. Such technology is capable of very high power, high efficiency HPAs (better than 5W/mm gatewidth in comparison

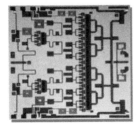 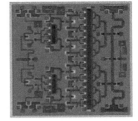

Figure 12.3 Examples of C and X-Band 10W HPAs fabricated with power PHEMT process

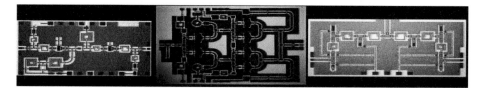

Figure 12.4 Examples of X-Band GaN Front-End chip-set

to 0.5 W/mm for GaAs PHEMT), robust LNAs, and high power switches, and it constitutes the basis for next-generation wideband multi-function multi-role T/R modules.

12.4.4 Photonics

Photonic technologies can be used in a large number of radar system applications grouped into two broad fields: microwave and digital photonics.

12.4.4.1 Microwave Photonics: Low Phase Noise Optoelectronic Oscillators

A photonic approach makes it possible to implement microwave signal generation by realising an Opto Electronic Oscillator (OEO) with significantly improved performance. In such an oscillator, the fibre optic loop is either based on a long fibre-based delay line or in an optical resonator (microsphere or microdisks). This beam is photodetected, filtered and amplified in the electrical domain and fed back to the external modulator. Once the overall loop gain of the optically-carried microwave signal is larger than one, then microwave oscillation takes place. Basically, the ratio between the loop length (with respect to the storage time in the loop) and the microwave wavelength (with respect to the period of the microwave signal) sizes the spectral purity of such an oscillator. The possibility of implementing quite long loops by using optical fibres, with very low optical losses (0.2 dB/km) makes this approach quite interesting for phase noise reduction.

12.4.4.2 Microwave Photonics: Optical Beam-Forming

Optical beam-forming for phased-array antennas offers many advantages such as small size, low weight, no susceptibility to electro-magnetic interference, wide instantaneous bandwidth or squint-free array steering.

The combination of large phase-steered arrays (for spatially-combined high power or for high angular resolution) and wide bandwidths (for large communication channels, high range resolution, spread spectrum, multipath mitigation, etc.) implies beam squint. The optical true time delay makes it possible to solve the squint problem of the large wide band array. The increasing use of photonics will enable the evolution of multifunction systems where radar, electronic warfare, and communications channels work together.

The architectures for signal distribution can be based on different multiplexing: WDM (Wavelength Domain Multiplexing) and TDM (Time Domain Multiplexing). WDM solutions usually employ optical sources able to generate multiple wavelengths such as arrays of DFB lasers, broadband source and fast tunable Laser. An alternative approach employs dispersive medium to realise delays such as Fibre Bragg Grating (Chirped) or High Chromatic Dispersion Optical Fibres.

12.4.4.3 Digital Photonics: Photonic A/D Conversion

Focusing on demultiplexing (or time-interleaved) photonic ADCs, typical optical architecture implies to: generate a stream of sampling optical pulses; modulate the height of the optical pulses by the voltage signal to be sampled through an optical modulator; split along multiple (N) parallel channels the samples to be A/D converted; perform A/D conversion on each channel with 1/N sampling rate using standard electronic A/DCs and finally recombine the bit stream by digital processing. The main advantage of a parallel approach in the optical domain is the availability of optical pulsed sources with very low jitter (<50fs) which can be used as a reference clock for all parallel channels. The distribution of such a clock may be done along the multiple channels with very low noise degradation.

12.5 Signal and Data Processing

12.5.1 Current Processing Technologies

The processing of radar systemic signals, imaging, and multi-domain sensors requires growing and enormous calculating power in order to obtain improved hard real-time performance as well as implementing knowledge-based algorithms.

At present we are extensively adopting Xilynx millions of gates "Field Programmable Gate Array" devices, G4 PowerPC and TigerSharc DSP in proprietary and commercial off-the-shelf circuits. Unfortunately the present steady growth of micro-processor calculation power and communication bandwidth performances (Moore's and MetCalve's Laws) are reaching a saturation point due to approaching limit on miniaturization, the widening gap between microprocessors and memory speeds, the diminishing performance returns resulting from clock frequency increases, and the increasing heat dissipation problems.

12.5.2 Multi-core Processors

In order to overcome this performance wall, microprocessor architects have realized that instead of going faster, a sensible approach was to use the chip area to exploit coarser parallelism rather than what was already provided by instruction level parallelism and thread level parallelism. This has led to an architectural innovation in modern processor architectures which has resulted in the creation of multi-core and multi-threading processors. Multi-core architectures are creating the potential for a leap forward in the processing capability made available by a single chip. This has a great potential for computationally hungry applications such as radar processing, image processing, computer vision, etc., for obtaining unprecedented real-time performance as well as implementing more elaborate algorithms.

Some of the available products are Asymmetric Multi-Core Processors and Homogeneous MultiThreading Processors such as AMD Opteron, Intel Core 2, ... Dual Cores; Intel XEON X5355, Intel Core 2 Extreme QX6700 Quad Cores; tomorrow and the day after tomorrow, the SUN Niagara II & Rock family and INTEL Penryn & Nehalem family.

At present the IBM CELL BE family (in the 2009 CELL BE (1+8), eDP, 45 nm; or CELL BE (2+16) DCM seems to be the most suitable for state-of-the-art signal processing requiring algorithms.

The IBM CELL BroadBand Engine processor boasts nine processors on a die. It contains a Power Processing Element and eight Signal Processing Elements (SPE vector processors).

The computational performance is 205 Giga FLOP/s at 3.2 GHz; a high-speed data ring connects everything with a 205 Giga Bite/s maximum sustained bandwidth and the chip interface equals 25.6 Giga Bite/s. Main concerns are the software tools and the non real time Operative System. Architectural design and specific additional hardware and software have to provide concurrent multi-threading among multi-sensors & multi-channels.

CBE performances were evaluated defining as applications the Rotational Motion Compensation algorithm for airborne real-time Synthetic Aperture Radar imaging; and the Space Time Adaptive Processing that is the holy grail of radar analysts. Our experience has shown that CBE multi-core processor can provide speedup of 20 and 30 times when compared with G4 Power Pc or TigerSharc DSPs, while at the same time allowing a reduction in volume and power roughly by an order of magnitude. Moreover these improvements (Fig 12.5), once the correct application partitioning is devised, are gained without too much programming effort, and in relatively little time.

After the SUN Sparc1 success, the "Homogeneous Multi-Threading" Niagara 2 Processors will be capable of 8 Cores running at 2 GigaHz and 64 Threads. Real time cryptography in back-ground, flexibility and programmability are the Niagara

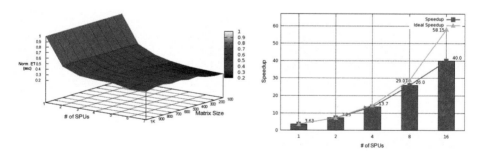

CBE / TigerSHARC speedup   STAP Normalized Execution Time
Figure 12.5 CBE Performance in rotational motion compensation and
the space time adaptive processing

II key points. In the future the Rock family will improve these features. In general, the main concerns are the latency control (not real time and deterministic) and the multi processes — multichip timing synchronization.

12.6 SW Enabling Technologies

Software technologies have a fundamental role for integration of sensors, systems and systems-of-systems. Software integration is strongly driven by the open architecture principles:

1. decoupling of functionalities
2. use of standard technologies
3. standardization of interfaces

Based on these principles, SELEX SI has developed several software architectural solutions that fulfil functional and performance requirements of its dual domain: civilian (Air Traffic Control, Homeland Security) and defence (C4ISR, Homeland Defence). Moreover SELEX SI is involved in major consortia for open architecture and software standardization; we are a contributing member of the *Object Management Group* (www.omg.org) and we attend the *Open Architecture Working Group (OAWG)* within the *Maritime Theatre Missile (MTMD) Defence Forum*. SELEX SI participates proactively in the following standardization processes: Radar Interfaces, Navigation Services, Meteorological Services and Middleware Services such as System Management, Enhanced View of Time, Load Balancing and others. Our software architectural solutions are the result of a continuous R&D process on the relevant and emerging technologies. The following paragraphs describe the leading software-enabling technologies.

12.6.1 Middleware

Middleware is the enabling technology of enterprise application integration. It describes a software item that connects two or more software applications so that they can exchange data. The software consists of a set of enabling services that allow multiple processes running on one or more machines to interact across a network. It is mainly used to support complex distributed applications such as Air Traffic Control and C4ISR systems. During the last ten years or so, SELEX SI has developed a strong competency in middleware products (either legacy and COTS) for mission critical systems with quality of services such as real time, fault tolerance, reliability and safety.

Moving from "systems" to "systems-of-systems" requires the evolution of "platform-centric middleware" towards "network-centric middleware". Because the networking infrastructure moves from LAN to WAN, the middleware software design has to deal with very different quality of service in terms of latency, scalability, timeliness, data persistence, time synchronization. For that reason, the SELEX

SI solution evolved from a platform-centric middleware, based on **CORBA** and **DDS** standard technologies, towards a network-centric middleware framework that integrates also **J2EE**, **SOA** and **mobile/wireless** software technologies.

12.6.1.1 CORBA & J2EE

CORBA (Common Object Request Broker Architecture) and J2EE (Java Platform Enterprise Edition) are types of middleware that allow applications to send objects and request services in an object-oriented system where application objects can reside either in the same address space (application) or remote address space (same host, or remote host on a network). Both systems enable independence from location and operating system. J2EE support only Java applications. As CORBA offers support for C++ and real time, it is widely used for mission critical systems development and integration.

12.6.1.2 Service Oriented Architecture

Service Oriented Architecture (SOA) is a paradigm for organizing and utilizing distributed capabilities that may be under the control of different ownership domains. It provides a uniform means to offer, discover, interact with and use capabilities to produce desired effects consistent with measurable preconditions and expectations. SOA represents a model in which functionality is split up into small, distinct units (services), which can be distributed over a network and can be combined together and reused to create business applications. These services communicate with each other by passing data from one service to another, or by coordinating an activity between one or more services. It is often seen as an evolution of distributed computing and modular programming. An SOA is commonly built using Web Service standards that have gained broad industry acceptance. These standards also provide greater interoperability and some protection from lock-in to proprietary vendor software. One can, however, implement SOA using any service-based technology, such as CORBA.

12.6.1.3 Data Distribution Service

The Data Distribution Service (DDS) is a "data-centric" middleware based on fully decoupling data producers and data consumers, and on the quality of service management. It has been designed for mission critical distributed application and standardized by OMG. SELEX SI is leading research activities on DDS, in collaboration with Information Systems Departments of Rome and Naples Universities, to optimize its application in a network centric environment.

12.6.2 Geographic Information System

A Geographic Information System (GIS) is a system for capturing, storing, analyzing and managing data and associated attributes which are spatially referenced to the Earth. GIS technology can be used for scientific investigations, resources management, asset management, environmental impact assessment, urban planning, cartography, criminology, history, sales, marketing and logistics. For example, GIS might allow emergency planners to easily calculate emergency response times in the event of a natural disaster, it might be used to find wetlands that need protection from pollution, or can be used by a company to site a new business to take advantage of a previously underserved market. GIS products are widely used throughout the warfare sector, as well as business and strategic domains. Many GIS products today offer web services interfaces to provide cartography and data exploitation via SOA. A standardization process of GIS open interfaces and protocols is managed by the Open Geospatial Consortium; OpenGIS Specifications support interoperable solutions that "geo-enable" the Web, wireless and location-based services. SELEX SI provides an architectural solution for seamless integration of GIS sold by different vendors. This solution maximizes the application design portability, allows multi-GIS integration and avoids lock-in with vendor.

12.6.3 SW Agents

SW agents technology seems to be very promising for the integration of networked heterogeneous systems. Agents are defined as autonomous, problem-solving computational entities capable of effective operation in dynamic and open environments, that acts on behalf of human, organization or other agents. Example of SW agents are: Buyer agents, User agents, Monitor-and-Surveillance (predictive) agents, Data Mining agents. They are often deployed in environments in which they interact, and possibly cooperate, with other agents (including both people and software) that may have conflicting aims. Multi agent systems require a semantic layer on top of the communication infrastructure in order to communicate and cooperate. They model the domain knowledge by means of ontology technologies. SW agents can be viewed as the future of Web Services: the "Semantic Web". SELEX SI is leading research activities on SW agents application, in collaboration with the Robotics Department of Rome University "La Sapienza".

12.7 Emerging Technologies

12.7.1 Revolutionary Technologies

The trend and roadmap of the sensors, and of the new-generation systems require a continuous technology innovation; however a major effort is required to maintain the mature technologies at the state of the art. Therefore, we are focusing our attention on the photonic and nanotechnologies innovative technologies because, compared to the evolutionary ones, they have a higher ratio between the results obtained and the effort required (utility function). These emerging, innovative and enabling technologies improve weight, size, speed, power consumption, efficiency, and so on. In addition they enable new solutions.

Innovative or revolutionary technologies could open new frontiers, since they are "killer technologies" requiring "creative destruction" of established technologies, thereby shifting capital from declining, mature technologies into those at the cutting edge (as clearly illustrated by Alan Greenspan).

12.7.2 Photonics Active Boards

The need for increasingly higher bandwidth and interconnect density pushes technology towards optics for the next generation of large scale integration devices. Electrical printed circuit boards will be replaced with waveguide based ones. Presently the main bottleneck for the use of Optical Boards (OBs) is the complexity of the production processes involved and the constraints imposed by the available materials. Therefore, OBs are confined to a prototype level with low active/passive element integration. Current R&D activities in SELEX SI aim at the development of new materials and processes to be used in active OBs fabrication.

The main challenges are: modelling and synthesizing organic molecules with smart optical properties, including nonlinear, sensor-functionalized and photovoltaic ones; developing structural materials (polymers and hybrid solgel) for organic waveguides; integrating passive (waveguides and connectors) and active (EO-modulators, switches, etc.) elements on the same board.

Critical issues to be addressed are managing optical losses on large-area devices by developing more advanced materials/processing schemes with cost reduction. This includes extensive exploiting of lithography (down to nanoscale) techniques with special emphasis on the combination of sub-wavelength optical lithography with imprinting. Relevant advantages are expected once the above mentioned items are overcome, in terms of enhanced data rate due to the large bandwidth (THz) and use of time and wavelength multiplexing; immunity to electromagnetic interference; higher interconnect density for intrinsic dimensions of optical patterns; low Non-Recurring Expenses (NRE) and unitary costs.

12.7.3 Photonics Technologies for Structural Health Monitoring

The main objective is the study and development of localised and distributed sensors for monitoring railway infrastructures, civilian building structures, tunnels, gas and oil pipes.

In particular we are considering the following types of sensors:

1. Distributed sensors based on Brillouin scattering for stress and temperature measurements
2. Distributed sensors based on Raman scattering for temperature measurements
3. Localised sensors based on Fibre Bragg Gratings for temperature measurements
4. Distributed sensors based on the analysis of the polarisation state of backscattered light scattering for temperature measurements.

12.7.4 Nano-Science and Nanotechnologies

Nanotechnologies represent an emerging domain with great potential. Nano is the creation of nanostructures, functional materials, devices and components, through the control of the matter on the nanometer scale of length, that is ten to the power of minus nine (10^{-9}) meters.

The "nano-scale" behaviour (size confinement and quantum mechanics) enables "orders of magnitude" improvements in weight, size, speed, power consumption, thermal management, strain, total efficiency, stealth, etc. Nano-science and nanotechnologies, as stated in Lisbon 2000, "could open a new era, enable, support and drive the 21st century knowledge-based society". Nano-science, the new theoretical and descriptive domain, from which different nanotechnologies are derived, is rightly defined as a "crucial, horizontal, qualifying science". It makes it possible to combine scientific disciplines that have been wrongly considered separated and different in the past.

Nanotechnologies, profiting by interdisciplinary and converging approaches, will contribute to the solution of problems that are typical of modern society. They will probably give a contribution to medical applications, and to research fields related to food, water, environment, energy production, creation of meta-materials. Nanotechnologies provides optimal performance regarding prognostics, photonics, information technologies (also through organic and inorganic nano-devices), biology, and in the science of cognition.

There is a flurry of activities in this field due to its attainable results, exceptional performance and revolutionary applicability. The importance and potentiality are approved by the European Commission with funds of over 3.5 billion euros in the 7th Framework Program.

At present, the main "nano" achievements are: more than four hundred "nano-enabled" products; more than 100,000 "workers"; more than ten Nobel's prizes: worldwide annual investments greater than 12 billion euros; more than 20,000 papers published with more than twelve thousand patents filed per year. The 2015 market forecast exceeds one trillion dollars!

12.7.5 Nanotechnologies in SELEX SI

SELEX SI intends to face the evolution of radar, system, and system of systems as a leader by proactively supporting and developing state-of-the-art enabling and innovative technologies. To do so, a century after the first RADAR "idea", we registered the NODAR (Nanotechnology Optical Detection And Ranging) trademark. NODAR ® aims to implement multifunctional, multi-role, multi-domain sensors, as well as adaptive, flexible, knowledge-based sensors, by including both photonic and nanotechnologies.

The "NODAR" technologies aims to implement multifunctional, multi-role, multi-domain sensors, as well as adaptive, flexible, knowledge-based sensors.

In addition to the photonics technologies described above, we identified and we are proactively developing in collaboration with many universities and research centres:

– Nano-structured frequency selective nano-materials for stealth and radar radomes.
– Carbon nano-tubes (CNT) enabling thermal management and interconnection". Properly dispersing 1% of SWCNT in our epoxy silver-loaded soldering glue Epotec H20E, we obtained a 30% reduction in thermal resistance.
– Nano X-ray source, using tungsten tip plus carbon nano-tubes in order to achieve very small size, weight, power consumption and spot size in presence of very high efficiency, X-ray density, portability and lifecycle.
– Nano-sensors using CNT since they have the highest equivalent surface, ambient temperature operation, fast response and recovery times. To obtain the agents identification, we use different types of sensors such as multi-finger electrodes with differently functionalized CNT plus quartz crystal nano-balances. The agents detection sensitivity is superb, while unfortunately the problem all over the world is agents identification!
– Carbon monoxide nano-photonic sensors enabled by metal-porphyrines.
– Vacuum tube nano amplifier using several clusters containing millions of carbon nano-tubes (as a 21st century cathode) since CNTs are highly efficient, low-temperature field emission sources.

We obtained very interesting preliminary results and our "reverse nano-triode" is working properly. We hope to use our nano-valve for TeraHz generation since it could theoretically amplify with high gain up to 37 TeraHz. Other partners are working on an Optical TeraHertz Source 7th FP project using Quantum Cascade Lasers technologies. TeraHertz technologies are very important for health, space communications and security applications.

The EU Commission has stated that the transformation of industries requires a truly integrated approach, either "vertical", combining materials sciences, nanotechnologies and production technologies, as well as information technologies or biotechnologies; or "horizontal", combining multi-sectoral interests. Multiscale is a unified strategy to link the customer operative requirements with innovative high-tech products and systems, with the basic researches, with technology developments, engineering and manufacturing processes. A "two-way" integration between science and engineering is required.

Four main Finmeccanica companies, SELEX SI, SELEX Communications, Alenia Aeronautica and CSM, launched the Finmeccanica Initiative named "Nanotechnology Multiscale Project" (NMP) in collaboration with at least ten universities and research centres. The project's main requirement is the definition and the development of integrated methodologies and environments to study, design, develop and test nanotechnology-based meta-materials, devices, sensors and systems. In addition, the project will closely investigate the environmental impact of nano-materials and nanotechnologies.

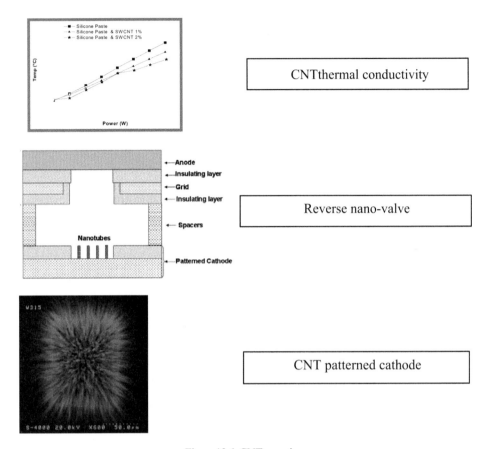

Figure 12.6 CNT overview

Our goal is to move directly from user requirements to products, research and functionality and finally to basic nano-materials production. And vice versa, to move from the theoretical description of the nano-materials domain to micro, to macro and finally to user performance! The Finmeccanica Nanotechnology Multiscale Project (NMP) enables Group companies to achieve industrial goals. It is the realization of vertical and horizontal integration of nano-science and engineering, recognized by the EU as a condition for nanotechnology applications for industry and society.

Chapter 13

Optimization of Flower Constellations for Dual Use

Daniele Mortari

Aerospace Engineering, Texas A&M University, College Station, USA

13.1 Introduction

The theory of Flower Constellations (FCs) constitutes a new methodology discovered and proposed by Texas A&M University to design satellite constellations. This chapter is meant to give a brief overview of present status of the FCs theory and few brief insights on some potential applications for Dual use. FCs have been discovered as a natural consequence (and extension to *n*-satellites) of the theory of *compatible* orbits (also called *resonant* or *repeating ground track*). Compatible orbits constitute a set of special orbits whose orbital period is synchronized with the period of a rotating frame. Obviously, the Earth-Fixed Earth-Centered (ECEF) rotating frame is particularly (or *the* most) important. However, the "compatibility" concept is a relative concept (relative to a rotating frame); hence any orbit can be seen as compatible. In particular, any orbit is also compatible with an infinite set of rotating reference frames.

An important aspect of a satellite on a compatible orbit is that its trajectory in the rotating reference frame constitutes a closed loop with assigned repetition time. Compatible orbits can be easily built upon the assumption of axial-symmetric field force model. This would include all zonal harmonics of Earth's gravitational field. However, compatible orbits are still possible when including the complete Earth gravitational model (Refs. 11–12).

The FCs theory explains how to place satellites on the same relative trajectory. This way the whole constellation is made of satellites that are running along relative trajectory one after another. Figures 13.1 and 13.2 show two examples of relative trajectories where 30-satellite constellations have been placed.

The FCs theory yield many by-products. The very first of them was the *Harmonic Flower Constellations* (or *Secondary Paths*). These kinds of constellations constitute new space objects. The whole constellation forms a rigid object whose shape is preserved during the repetition time. This new space object spins about the constellation axis which, in turn can be freely oriented in space. Just as examples, Figures 13.3 and 13.4 show the "shield" and "elix" harmonic (shape-preserving) FC. These new space *objects* are obtained when the design parameters satisfy some specific conditions (Ref. 8). In general, their shapes are unexpected and non-intuitive solutions that cannot be derived from the relative trajectory.

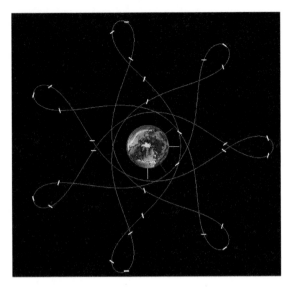

Figure 13.1 $N_p = 7$, $N_d = 3$, $i = 0$, $e = 0.584$

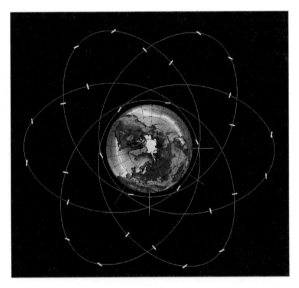

Figure 13.2 $N_p = 6$, $N_d = 1$, $i = 0$, $e = 0.454$

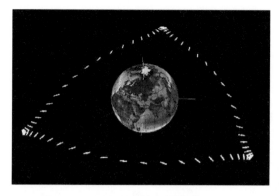

Figure 13.3 "shield" FC

Figure 13.4 "Elix" FC

Other by-products are the "*dual*" FCs (Ref. 10) where the satellites are in orbits compatible with two rotating reference frames. Reference 10 shows how to design dual FC simultaneously synchronized for Earth & Mars and for Earth & Jupiter systems. These constellations are proposed to design interplanetary communication networks to support future missions to Mars and Jupiter, respectively. Important by-product are the *two-way* FCs, useful for Earth observation as well as for Reconnaissance systems. More details on these constellations will be given on the regional coverage section.

Table 13.1 summarizes various aspects of Flower and Walker constellations design methodologies. Since Walkers use circular orbits only, the perigee control problem, due to the apsidal line rotation, is avoided. However, constraining the solution using circular orbits only the space of potential solutions is dramatically impaired.

Station-keeping costs to compensate gravitational and non-gravitational perturbations affect, in general, Walkers and Flowers with no preferences. On the other hand, Flowers require axial-symmetric force model (about the pole). In first approximation, the Earth gravitational field well satisfies this requirement. Obviously, drag, third-body effects, and/or non-zonal harmonics, must be compensated.

Table 13.1 Comparison table between Walker and Flower Constellations

Type of orbit or characteristic	Flower	Walker
Circular Orbits	Yes	Yes
Elliptical Orbits	Yes	No
Free choice of inclination	Yes	Yes
Choice of revisiting time with multiple satellites	Yes	No
Repeating ground track	If desired	If desired
"Two-way" orbits (Ref. 2)	Yes	No
"Dual Compatible" orbits (Ref. 10)	Yes	No
Multi-stationary Orbits (e.g. Molniya)	Yes	No
Sun Synchronous	Yes	Yes
Arbitrary number of satellites	Yes	No
"Harmonic" constellations (Figures 1 and 2)	Yes	No

The axial-symmetric force model assumption is still satisfied if the full-zonal harmonics are considered. These other perturbations affect any constellation built on symmetry or axial-symmetry based criteria.

13.2 Optimization Techniques

The design parameters of a N-satellite FC with specified initial time are the following:

1. Three parameters ruling the compatible orbit
   a. Angular velocity of the compatible rotating reference frame ($\omega_\otimes$). If $\omega_\otimes$ is selected as the Earth spin rate, then the satellites of the FC are on the same repeating ground-track relative trajectory.
   b. Number of "petals" (N_p) and number of "days" (N_p). These two integers are ruling the orbital period (and semi-major axis). They represent the number of orbits and the number of rotations of the compatible rotating reference frame during the constellation repetition time, respectively.
2. Three integer parameters ruling the phasing (F_n, F_d, and F_h). These integers are responsible for the distribution of the satellites in the admissible locations.
3. Three orbital parameters for space orientation and shape. They are the perigee altitude (or orbit eccentricity), the orbit inclination, and the argument of perigee.
4. The initial values (two parameters) of right ascension of ascending node and mean anomaly of the first satellite.
5. Two constellation orientation parameters. These are the two angles defining the direction of the FC axis in space.

Since the design parameter set involves several integers, analytical optimization techniques can be applied only when these integer parameters are defined. To find the integer design parameters, Evolutionary Algorithms (EAs), such as Genetic Algorithms or Particle Swarm Optimization technique, have been successfully used. In general EAs do not guarantee the finding of *the* optimal solution. However, the capability of finding multiple quasi-optimal solutions makes them quite attractive. It is, indeed, impossible to include in a cost function all the optimization criteria and all the constraints. Therefore, having various quasi-optimal solutions allows us to grade them for all the other aspects not included in the cost function, such as orbit maintenance costs, launch costs, radiation levels (Van Allen belts), constellation deployment time and costs, etc.

Flower Constellations have been optimized for Global Navigation Systems (Refs. 13–14) and for interferometric imaging systems (Refs. 15–16), where the "Flower Formations Flying" have been introduced to design satellite formations with prescribed limited distances. The Flower Formations Flying, which are not derived from the compatible orbit theory, uses the FCs phasing mechanism. These formations are built using a circular or an elliptical reference orbit.

13.3 Dual Use Applications

The term "dual-use" can be intended as military-*and*-civil applications or as multipurpose applications such as communications and Earth observation. Independently from the original meaning, such systems require judicious displacement of constellation satellites such that a weighted index of merit, including costs for both applications, is minimized. Because of the many optimization criteria to satisfy the problem becomes pareto-optimal. Therefore, the solution is then selected (among many potential configurations) based on robustness, reliability, costs, future plan, and experience-derived considerations.

Obviously, a dual-use mission implies there are common aspects between the two distinct applications: the type of orbit (LEO, MEO, HEO, GEO) and the satellite stabilization must be, anyway, identical for both applications. Therefore, rather than define specific dual-use missions, this section faces the dual-use FC optimization problem by finding the orbital characteristics and the constellation dynamics that can be considered "optimal" for more than one space missions. For this reason in the next two sections we consider dual-use FCs two different kind of constellations, those whose satellites remain uniformly distributed in space (space-uniform) and those characterized by satellites uniformly distributed in time along the relative trajectory (time-uniform).

13.4 Flower Constellations Uniformly Distributed in Space

The property of having the constellation satellites distributed over the celestial sphere as uniformly as possible over the whole repetition time often coincides with

Figure 13.5a 10-sats space-uniform prograde Flower Constellation

Figure 13.5b 10-sats space-uniform retrograde Flower Constellation

the optimality definition for some applications. Typical examples are global communications and Earth observation systems.

Flower Constellations can be designed using EAs with satellites distributed as uniform as possible over the celestial sphere. This is done using the relaxation approach to the Thompson problem (Ref. 9). This approach considers the satellites as electrostatically charged. The optimality is obtained when the maximum value of the overall potential energy during the repetition time is minimized. Two examples of prograde and retrograde 10-satellite space-uniform FCs are given in Figures 13.5a and 13.5b.

13.5 Flower Constellations Uniformly Distributed in Time

In the Flower Constellation theory, the position of the k-th satellite along the relative trajectory is identified by a new mean anomaly (Γ_k) that can be derived (Ref. 8) from the compatible parameters (N_p, N_d) and the FCs initial phasing (M_k, Ω_k). This is obtained by solving a Diophantine equation and by a reconstruction technique known as the "Cinese Remainder Theorem". The mean anomaly Γ_k plays an important role in designing FCs with satellites uniformly distributed along the relative trajectory because this kind of distribution is optimal for Earth observation systems and for reconnaissance missions to observe specific sites/regions. As an example, a 7-sats time-uniform prograde Flower Constellation is shown in Figure 13.6. The green area represents the overall coverage for the sensor field of view considered while the red areas are the instantaneous observed regions. This FC has a repetition time of one day. Therefore, because of the time-uniform distribution, the constellation completes the overall coverage seven times per day. In particular, the time

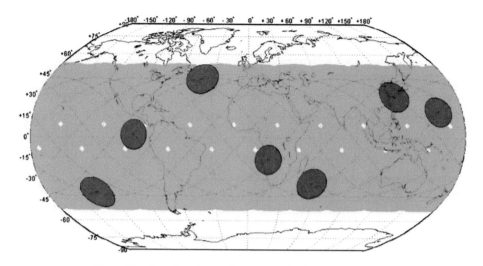

Figure 13.6 7-satellite time-uniform prograde Flower Constellation

interval between two subsequent satellite passages, as seen from any point of the ground track, is constant (time-uniform).

13.6 Flower Constellations for Coverage and Visibility

Flower Constellations can be specialized for regional coverage. Civilian and military applications are here characterized with different requirements and optimality definitions. Usually, the mission requirements belong to one of the following:

1. Continuous observation of given geographical regions (and/or Earth sites) with assigned number of simultaneous observing satellites.
2. Observation of given geographical regions (and/or Earth sites) with prescribed frequency of observation and using a minimum number of satellites.

The above mission requirements are easily and immediately satisfied using FCs because the problem consists of 1) finding the optimal compatible orbit — and the FC are built on compatible orbits — and 2) uniformly distribute the satellites along the relative trajectory, an easy task using FCs.

Obviously, the optimization can be constrained by one or more of the following items: a) avoid Van Allen belts, b) bounded costs of orbit maintenance (unfortunately, this aspect deeply affect using elliptical orbits), c) launch costs, d) bounded costs for augmented systems reconfiguration and/or for reconfiguration due to single failure, and e) minimum acceptable resolution. In general, LEO orbits allows to achieve the best resolution but LEO orbits are also perturbed by gravity gradient and drag (FCs are J_2 compliant).

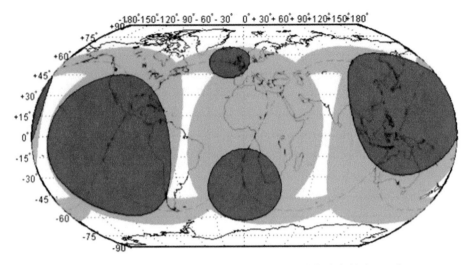

Figure 13.7 "Two-way" Flower Constellation for regional/site hybrid observations

An interesting (and unique) capability for coverage and visibility is offered by the two-way orbits and associated two-way FCs (see Ref. 2). These orbits are characterized by a ground track path that is a closed-loop trajectory intersecting itself, in some points, with tangent intersections (see Figure 13.7). The spacecraft passes over these tangent intersections once in a prograde and once in a retrograde mode. Earth observation systems and Reconnaissance systems can take advantage of these propriety by having simultaneous observations of the same target area from different altitudes (sensors).

Acknowledgments

The author would like to thank my former PhD students, Dr. Wilkins and Dr. Bruccoleri, for being the two leading columns in the FCs discovery. In particular, Dr. Wilkins has been instrumental in the theory development (Refs. 5–6, 17) while Dr. Bruccoleri's contributions (Refs. 14, 18) were in optimization and in developing "Flower Constellation Visualization and Analysis Tool" (FCVAT). Without FCVAT it would be practically impossible to take the FCs at today's stage of maturity. Initial contributions were also given by my former Ph.D. students, Dr. Park and Dr. Abdelkhalik, to design FCs for Global Navigation System (Ref. 13), for ground reconnaissance problem from space (Ref. 4), and for the theory of the two-way orbits (Ref. 2). I want to express my gratitude to Dr. Tonetti, who shared the development of the "Flower Formation Flying" (Refs. 15–16), and to Dr. Avendano, for bringing the FC theory to a deeper level of maturity (Ref. 8). Finally, I would like to express my gratitude to Prof. Ruggieri (Tor Vergata Univ.) for leading a group of researchers who performed a thorough analysis on theory and potential applications and to Dr. Izzo (ESA-ESTEC) for excellently over viewing the research contract (Ref. 7).

References

[1]   Mortari, D., Wilkins, M.P., and Bruccoleri, C. "The Flower Constellations," *Journal of the Astro-nautical Sciences*, Special Issue: The John L. Junkins Astrodynamics Symposium, Vol. 52, Nos. 1 and 2, January–June 2004, pp. 107–127.

[2]   Abdelkhalik O. and Mortari, D. "Two-Way Orbits," *Celestial Mechanics and Dynamical Astronomy*, Vol. 94, No. 4, April 2006, pp. 399–410.

[3]   Mortari, D. "Flower Constellation as Rigid Object in Space," *ACTA Futura*, Issue 2, August 2006, pp. 7–22.

[4]   Abdelkhalik O. and Mortari, D. "Orbit Design for Ground Surveillance Using Genetic Algorithms," *Journal of Guidance, Control, and Dynamics*, Vol. 29, No. 5, September–October 2006, pp. 1231–1235.

[5]   Mortari, D. and Wilkins, M.P. "The Flower Constellation Set Theory. Part I: Compatibility and Phasing," *Transactions on Aerospace and Electronic Systems*. To appear in April 2008.

[6]   Wilkins, M.P. and Mortari, D. "The Flower Constellation Set Theory. Part II: Secondary Paths and Equivalency," *Transactions on Aerospace and Electronic Systems*. To appear in April 2008.

[7]   Ruggieri, M., De Sanctis, M., Rossi, T., Lucente, M., Mortari, D., Bruccoleri, C., Salvini, P., and Nicolai, V. "The Flower Constellation Set and its Possible Applications," ESA-ESTEC, Ariadna ID: 05/4108, Contract No. 19700/06/NL/HE. Final Report, June 16, 2006.

[8]   Avendano, M. and Mortari, D. "Rotating Symmetries in Space: The Flower Constellations," In progress.

[9]   Saff, E.B., and Kuijlaars, A.B.J. "*Distributing Many Points on a Sphere*," *Mathematical Intelligencer*, Vol. 19, No. 1, 1997, pp. 5–11.

[10] Mortari, D. and Wilkins, M.P. "Dual-Compatible Flower Constellations," Paper AAS 06-202 of 2006 Space Flight Mechanics Meeting, Tampa, Florida, January 22–26, 2006.

[11] Russell, R.P. and Lara, M. "Long-Lifetime Lunar Repeat Ground Track Orbits," *Journal of Guidance, Control, and Dynamics*, Vol. 30, No. 4, July–August 2007, pp. 982–993.

[12] Russell, R.P. and Brinckerhoff, A.T. "Eccentric Orbits around Planetary Moons," Paper AAS 08-181 of 2008 Space Flight Mechanics Meeting, Galveston, Texas, January 27–31, 2008.

[13] Park, K., Wilkins, M., Bruccoleri, C., and Mortari, D. "Uniformly Distributed Flower Constellation Design Study for Global Positioning System," Paper AAS 04-297 of the 2004 Space Flight Mechanics Meeting, Maui, Hawaii, February 9–13, 2004.

[14] Bruccoleri, C. "Flower Constellation Optimization and Implementation," PhD dissertation, Texas A&M University, Aerospace Engineering, December 2007.

[15] Mortari, D. and Tonetti, S. "The Flower Formation Flying. Part I: Theory," Paper 08-180, 2008 AAS Space Flight Mechanics Meeting, Galveston, Texas, January 27–31, 2008.

[16] Tonetti, S., Hyland, D., and Mortari, D. "The Flower Formation Flying. Part II: Application," Paper 08-186, 2008 AAS Space Flight Mechanics Meeting, Galveston, Texas, January 27–31, 2008.

[17] Wilkins, M.P. "The Flower Constellations Theory, Design Process and Applications," PhD dissertation, Texas A & M University, Aerospace Engineering, December 2004.

[18] Bruccoleri, C. and Mortari, D. "The Flower Constellations Visualization and Analysis Tool," *2005 IEEE Aerospace Conference*, Big Sky, Montana, March 5–12, 2005.

[19] Park K., Ruggieri, M., and Mortari, D. "Comparisons Between GalileoSat and Global Navigation Flower Constellations," *2005 IEEE Aerospace Conference*, Big Sky, Montana, March 5–12, 2005.

Chapter 14

Dual Use Technology and Applications: the GALILEO case

Carlo des Dorides

GNSS Supervisory Authority, Bruxelles, Belgium

14.1 Introduction

The GALILEO Project, the European flagship project, is at the time this article is written (September 2007) in a delicate phase: following an intense tendering process and about two years of extensive negotiation, the project has recently faced the formal declaration to terminate the negotiation due to the lack of prospect to positive conclude it in a reasonable timeframe. Among other elements, the adopted contractual structure, namely the Public Private Partnership/Concession, has been charged as one of the main reason for such failure. The GALILEO Project is then facing a complex phase of restructuring. The article which follows will describe the process undertaken and provide some insights about the challenges which have been faced with the PPP/Concession approach together with some hints about the expected near term steps. Despite of the difficulties, the strong political will to make GALILEO happens, provides the ground for an optimistic view of the GALILEO future.

Specific emphasis will also be addressed about the intrinsic "dual use" nature of the system and the "Public Regulated Service" (PRS), one of the five service categories provided by GALILEO. The article will also point out the issues at stake concerning the policy to access PRS and its specific implementation process. Finally some preliminary assessment about the expected value of PRS services will be provided together with potential revenue generation mechanisms.

14.2 The GALILEO PPP/Concession Structure

GALILEO has been conceived as the first European PPP (Council Conclusions of 5th April 2001), assuming the Private Sector to:

- finance 2/3 of the project ($\sim$ 3.0 b€)
- operate & maintain the system for the overall concession lifetime (20 yrs)
- replenish & exploit the system.

Remuneration of the injected capital investments and the repayment of debt was supposed to come from revenue generation. Value-for-money for the Public Sector implied effective risk transfer i.e. risk should be retained by the party best capable to manage it.

The PPP structure, seen as a possible approach for the space sector, showed some benefits by:

- encouraging competitive market position of industry in accessible engineering market;
- promoting research and development with new space applications with a potential spin-off into the commercial sector;
- by creating synergies between public and private services and needs, especially for the public regulated services (governmental and institutional markets);
- by putting pressure for return on investment to better focus user and market needs.

14.3 GALILEO: The Current Situation

Entrusted by the March 2007 Transport Council EC/ESA/EIB/GSA have:

- assessed the ongoing negotiation process
- analysed the margin on successful completion
- evaluated possible alternative scenario, including those which assume a more substantial involvement of the Public sector in the system development

Following the assessment of the ongoing negotiation process and the analysis on the margin of its successful completion, in June 2007 the GSA decided to terminate the GALILEO tender Procedure. A number of different reasons contributed to the failure of the initial GALILEO PPP procurement process and the difficulties which were encountered by the public sector in its negotiations with the Merged Consortium.

These difficulties included, in particular:

a) the technical complexity of EGNOS and GALILEO;
b) disputes and disagreements regarding industrial governance and work share;
c) the inability of the public sector to transfer certain risks (including in particular, market risk and design risk) to the private sector.

14.3.1 Technical Complexity of EGNOS and GALILEO

One of the most significant problems which affected the commercial negotiations between the public sector and the private sector during the initial PPP procurement process concerned the technical complexity and the untested and unproven nature of much of the technology on which GALILEO relies. It is clear that this, together with the lack of reliable test data for some of the GALILEO system components (and the GALILEO system as a whole), was a major difficulty for the Merged Consortium in assessing the technical risks associated with the GALILEO Project.

14.3.2 Industrial Governance and Work Share

Another important issue which affected the negotiations between the public sector and the Merged Consortium in the latter stages of the initial procurement concerned the apparent lack of agreement between the Merged Consortium's member companies regarding the internal division of the relevant GALILEO infrastructure and the work share for each company involved. This lack of agreement had a profound effect on the Merged Consortium's ability and willingness to progress the negotiations with the public sector.

A related difficulty encountered by the public sector during the initial procurement process stemmed from the apparent lack of agreement between the members of the Merged Consortium in relation to key risk issues (including market risk and design risk) and the inflexible decision-making process within the Merged Consortium which required the unanimous agreement of all of the member companies for all commercially significant decisions.

14.3.3 Difficulty of Transferring Certain Risks to the Private Sector

Another significant challenge which arose in the initial PPP procurement process was trying to agree an appropriate commercial and legal risk allocation position as between the public sector and the private sector. Because of the size and complexity of the GALILEO PPP, it was extremely difficult for the negotiating parties to agree a sensible risk allocation position which met the public sector's stated objectives for the GALILEO PPP.

14.3.3.1 The difficulty of transferring market risk

The Commission has acknowledged that the market for global satellite navigation services of the sort that would be provided by the GALILEO concession holder continues to be uncertain and that as a result of that uncertainty, the Merged Consortium expected the EU to underpin the associated market risks. The Commission concluded that this element of the PPP was underestimated in the original plans for GALILEO and that the public sector had always proceeded on the assumption that the private sector would be willing to assume market risk. However, achieving an effective transfer of market risk was one of the most difficult challenges faced by the public sector in its negotiations with the Merged Consortium.

14.3.3.2 The difficulty of transferring design risk

From the beginning of the initial PPP negotiations, one of the most important areas of commercial disagreement between the public sector and the private sector was the issue of technology and design risk. This risk gave rise to a number of difficulties. The difficulties were due not only to the technical complexity of the GALILEO design and the outputs to be achieved by the future GALILEO Operating Company (GOC) during the operational phase of the Project, but also the split in responsibility for design and development of the GALILEO system on the one hand and deployment, operation and maintenance of the system on the other.

Although the GOC was ultimately expected to be responsible for the operational performance of the GALILEO constellation under the initial PPP structure, ESA had always been responsible for developing the design of the satellites and other Project assets required for delivering IOV and for completing the necessary IOV tests. This split in responsibility caused significant difficulties for the private sector, which consequently refused to accept a full design risk transfer on these terms.

14.4 Way Forward of the GALILEO Project

The June transport Council required further analysis on the possible alternative scenario's which are currently ongoing, namely:

–   Deployment phase: financed entirely by the public sector up to Full Operation Capability (FOC — 30 satellites) or up to Initial Operational Capability (IOC — 18 satellites).
–   Operational/Maintenance/Exploitation phase: service concession or a service contract.

Supporting EC answer to the Council, GSA conceived different alternatives for structuring the future of GALILEO after Full Operation Capability — FOC along the following basic structures:

a)   Public sector delivers FOC+ PPP concession for operation and service provision
b)   Public sector delivers FOC+ service contract for operation and service provision
c)   Public sector delivers FOC+ separate service contract to FOC+[5] years, then PPP concession

14.5 The GALILEO Public Regulated Service

In all the scenarii outlined above the GALILEO system will continue to be a civil system for civil applications which can be also used for governmental applications through the PRS.

PRS is a secured service, encrypted and resistant to jamming and interference, reserved principally for the public authorities who demand a high level of continuity, especially in situation of crisis or presence of threat. PRS signals are separated from other GALILEO service signals and are also broadcasted in a dedicated frequency band.

The access to PRS is controlled for security grounds and therefore requires technical (e.g., cryptology), procedural (e.g., security accreditation) and potentially also decisional bodies and procedures to ensure the adequate control of the service and its users.

In addition, potentially, there are numerous PRS users with widely differing operational needs, behaviour and levels of confidence. The control of access to the PRS users therefore needs to be based on a policy of access.

14.5.1 *Objectives of the Policy of Access to PRS*

The objectives of the policy of access to PRS are the following:

- To organise the PRS users through a system providing efficient organisation which sustains the PRS security; in particular to define the conditions used by Member States to organise and manage their User Groups;
- To define the necessary interfaces between the infrastructure for decision making (Council, Member States, Commission) and the implementation (Supervisory Authority, concessionaire, manufacturers, users);
- To put in place the necessary framework for technical and decision making processes in time for the operational phase of GALILEO.

PRS Access Policy is expected to respect the following high level principles:

- All EU Member States will have unlimited and uninterrupted access to PRS which shall comply with specific security requirements
- Strict control over the PRS use will be implemented in order to guarantee that PRS is protected against unauthorised access
- PRS manufacturers will have to follow the specific rules in terms of certification and security accreditation
- Member States will define their national users organising them in national groups and will maintain an high level of control over them

The implementation of the policy of access to PRS will also comprise:

- Organisation of the PRS User Groups in such a way as to guarantee that when a PRS group is compromised, it shall not impact on the operational capacity of other PRS groups;
- Coordination of the various public and private actors by attributing to them clear and limited responsibilities, in order to ensure a secure and efficient management of PRS;

- Implement the Council Joint Action as concerns PRS usage and report the progress of the implementation of the Policy of access to PRS to the Council on a regular basis;
- Enable the Council to decide or reject potential demands by non EU Countries to use PRS and on the export of PRS equipments to such countries.

14.6 Value of Public Regulated Services

Main PRS application domains are expected to be:

- internal security and law enforcement
- custom
- critical transport, energy and telecom
- emergency services
- defence

In terms of revenue projections, recent study have shown that about 30% of the overall forecasted revenues could come from PRS. Revenue generation mechanisms are supposed to come from two main sources:

- PRS Signal in Space: Annual subscription fee per country (lump sum) on a non commercial basis
- PRS receivers: Receiver fee per unit or Receiver rental (annual contract with service providers — fee based on the volume PRS receiver rented throughout the year)

14.7 Conclusions

GALILEO system, though is a civil system under civil control, can be considered as one of the best example of dual technology, showing how such a characteristic can prove to be beneficial to the financial viability of the project.

Despite of the difficulties which the GALILEO Project has recently met, the political awareness and willingness to pursuit it is still high. At present the project is under a significant restructuring which entails the Public Sector being responsible of the full system deployment (Full Operation Capability) before the award of a contract for service provision. Final decision is expected by the Financial Minister and Transport Councils by year end.

Chapter 15

Italian Initiatives for Broadband Communications and Data Relay Satellite Systems

Mario Ciampini, Giacinto Losquadro

Thales Alenia Space Italia

15.1 Introduction

The "Italian Space Integrated Network for Security" (ISINS), represents an essential component to support the Network Centric Operations that need to be executed by the Civilian and Defence Institutions, in the various context, including interventions for Disaster Relief, Peace Keeping/Enforcing actions with Coalition/Joint Forces, etc. Transformation, interoperability and standardization are benchmarks for the SatCom Infrastructure evolution that should be fulfilled, starting from the existing Infrastructure (SICRAL System), through the deployment of the following two additional components:

a) the broadband satellite system, named FIDUS (i.e. the new Franco Italian Dual Use System);
b) the advanced Data Relay satellite System (DRS),

on which, today, is focused the attention of the Italian Space Agency, of the Defence/Internal Affairs Ministries and of the Italian Space Industry.

FIDUS is a EHF/Ka band satellite system, able to support the multiplicity of diversified broadband connectivity needs of the Italian governmental institutions. The FIDUS payload will be installed on a GEO medium class satellite (ATHENA-FIDUS satellite, a French-Italian Mission), providing high transmission capacity (about 2 Gbit/s); it will include:

- two bentpipe transponders: (i) Ka/Ka band and (ii) EHF/Ka band to support mainly star topology services on the Italian Coverage;
- a processed advanced EHF/Ka transponder operating on the Italian Coverage to support meshed connectivity services on the Italian Coverage;
- two forward and return bentpipe repeaters from/to two steerable coverage areas (theatres with 1500 Km diameter) interconnected with the Italian Coverage.

The DRS, together with the broadband satellite systems (that support terrestrial users), represents a fundamental space component, that will allow to meet all the SatCom requirements of LEO/MEO satellites, requiring worldwide connectivity, fast deployment, large transmission capacities, widespread dissemination of contextual information and finally an high resilience to damage of ground infrastructures. In particular the new Data Relay System (DRS) will represents the successor of the ESA-ARTEMIS satellite, extending the scope of the said system, serving also users of the Global Monitoring for Environment and Security (GMES) program, like aircrafts, unmanned aerial vehicles and land/marine vehicles. Therefore, the so defined Enhanced Data Relay Satellite Mission architecture would closely relate to a dual (or multi) use mission, mainly used for:

- ordering and collecting data to/ from Earth Observation (EO) satellites (including GMES and COSMO-SkyMed), UAV's, launchers, aircraft, maritime and land mobile vehicles;
- controlling the above platforms (TM-TC&RNG)to update their mission planning in such a quick reaction mode, that can be considered, when needed, real time or nearly-real time response.

This chapter, apart the summary description of the above mentioned components, in the concluding part, illustrates, the added value, for the Italian Society/Community, of these SatCom Infrastructures, when seen in the context of "System of Systems" matched to support NCO's.

15.2 Scenario

Nowadays, Homeland Defence and International Security cannot be separated, and must be guaranteed making full use of all the available resources, both military and civilian. In this context the dual-use characteristic of the SatCom assets becomes even more important.

Common challenges and objectives for the Security and Defence Communities are the new forms of interventions, new tasks, new assets and stringent reactivity constraints. The Italian SatCom Infrastructure represents indeed an essential component to support the Network Centric Operations that need to be executed by the Civilian and Defence Institutions, in the various contexts, including interventions for Disaster Relief, Peace Keeping actions with Coalition/Joint Forces, etc. Transformation, interoperability and standardization are benchmarks for the SatCom Infrastructure design. A very important evolution of such a space infrastructure is currently underway in this field, especially owing to the following factors:

a) the asserted need for the sharing of kind of information, specially needed to support Shared Situational Awareness (SSA), for the networking of all systems and all organizations, delivering significant benefits, especially by increasing efficiency and responsiveness (see Figure 15.1);

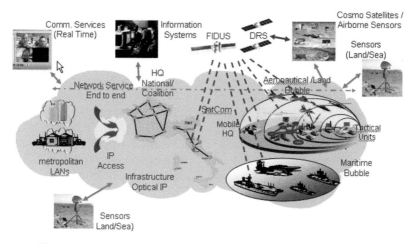

Figure 15.1 Broadband systems and data relay systems connectivity scenario

b)   the emergence of new IP-based technologies which progressively and broadly assert themselves through their capacity to deliver broadband access in an increasingly secured context;

c)   more generally the importance of innovation in the field of new services and solutions.

In this sense, although in a secure framework environment, three main requirements have to be highlighted: i) Large Bandwidth and Matched Service Areas, ii) Compatibility with evolving Standards and Protocol (i.e. IPv6), iii) Node switching and flexible connectivity.

These requirements, as seen in the current Italian SatCom context, should be fulfilled thanks to the evolution of the existing and operational Infrastructure (i.e. the SICRAL System), through the following two additional components:

d)   a broadband satellite system, named FIDUS (i.e. Franco Italian Dual Use System);

e)   an advanced Data Relay satellite System (DRS),

on which, today, is focused the attention of the Italian Space Agency, of the Defence and Internal Affairs Ministries and of the Italian Space Industry.

This evolving SatCom Infrastructure is providing and will further provide a major added value for the Italian Society/Community when seen in the context of "System of Systems" supporting NCO.

15.3 Broadband Satellite Communication: the FIDUS Initiative

At the present, satellite communications rely mainly on conventional/transparent repeater configurations, but the service and system requirements, for the near future scenario, lead toward "processed" systems: indeed the planned FIDUS satellite system (of the French-Italian ATHENA-FIDUS mission) will offer advanced on board processing capabilities for broadband and high quality meshed connectivity services.

The French and Italian Governments with their respective Space Agencies, CNES and ASI, and the Ministries of Defence, have expressed strong interest in space-born systems exploiting multimedia technologies and associated broadband capacities. With the dedicated dual solution of ATHENA-FIDUS System, Defence and other Governmental institutions would benefit from broadband multimedia services with a high capacity and low cost and user friendly terminals.

The system efficiency is optimised by using the civilian best space telecommunications standards, namely DVB-RCS and DVB-S2 (ETSI std, but also a proposed NATO 4622 Stanag). The main objectives of the mission are the following:

- provide communications for dual-use complementary to military existing SICRAL system;
- use of civilian broadband technologies in Ka and EHF bands in order to reduce significantly the cost of in orbit bandwidth versus actual military or commercial solutions;
- service start in the Y2011.

FIDUS will be implemented on a geostationary medium class satellite, providing high transmission capacity (overall throughput 2 Gbit/s). The payload, supporting the Italian mission, will include:

- two bentpipe transponders: (i) Ka/Ka band and (ii) EHF/Ka band to support mainly star topology services on the Italian Coverage;
- processed EHF/Ka transponder on the Italian Coverage to support meshed connectivity services on the Italian Coverage (see Figure 15.2a);

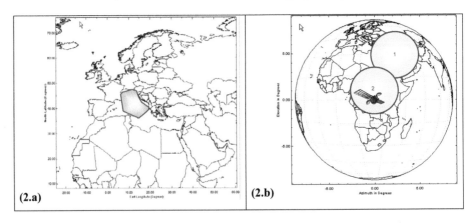

Figure 15.2 FIDUS Service Areas [fixed (a) and steerable (b) service areas]

– two forward and return bentpipe repeaters from/to two steerable coverage areas (theatres with 1500 Km diameter, see Figure 15.2b) interconnected with the Italian coverage.

Figure 15.3 highlights the following three main network segments:

– National Star Network (Hub, Satellite Terminal) on bentpipe EHF and Ka Band transponders;
– National Meshed Network (NOC, Satellite Terminal) on OBP EHF and Ka Band transponders;
– Star Theatres Network (Hub, Satellite Terminal) on bentpipe Ka Band transponders,

which support the following services:

– Broadband access network with star topology on Italian landmass territories and neighbouring areas;
– Broadband network with mesh topology on Italian territory and neighbouring areas, in single hop, minimum latency for real-time traffic exchange/small aperture terminals;
– Broadband access network on foreign theatres, for high rate broadcasting network on theatres and operational links between theatre users;
– Unmanned aerial vehicles relay between Italian landmass territories and Italian foreign theatres;
– Broadcasting of remote sensing images or data from Italian territories to Italian foreign theatres.

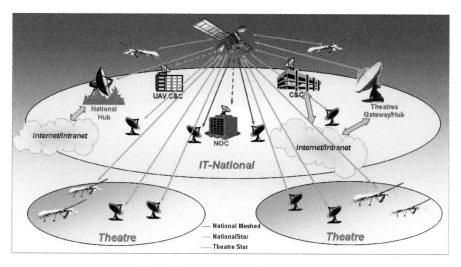

Figure 15.3 FIDUS Network Architecture

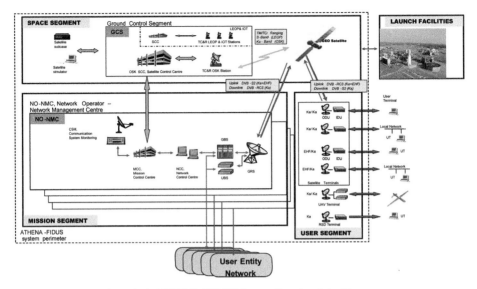

Figure 15.4 ATHENA-FIDUS System Functional Architecture

FIDUS system architecture is illustrated in Figure 15.3 with the Space Segment (Italian component of the ATHENA-FIDUS satellite, the Mission Segment (the Italian component of the French-Italian Mission & Control Segment) and the User Segment.

15.4 Data Relay Satellite System

The DRS, together with broadband satellite systems (that support terrestrial users), represents a fundamental space component, that will allow to meet all the Sat-Com requirements of LEO/MEO satellites, requiring worldwide connectivity, fast deployment, large transmission capacities, widespread dissemination of contextual information and finally an high resilience to damage of ground infrastructures.

Significant improvements to the Earth Observation missions in terms of enhancement of images throughput and performance can be achieved by a Data Relay and Tracking Geostationary Satellite System in terms of the ability to track and receive data from a number of Earth Observations LEO satellites and, in perspective, extended user types (such as UAV), as the result of the longer visibility period between a LEO and a GEO spacecraft (about 60 min. of a typical 100 min. LEO orbit) than that between a LEO and a ground station (about 10 minutes per orbit). This has been already validated by the ESA ARTEMIS Satellite, during huge and still ongoing acquisition campaign with ENVISAT and following the successful experimental trial campaign with SPOT IV via the SILEX optical data relay system. Overall, these perspectives fall significantly in the scope of a wide action plan for the above mentioned "Italian Space Integrated Network for Security" (ISINS), which the Italian component of the DRS will belong to. In particular the proposed Data Relay System (DRS) will represent the successor of the ESA-ARTEMIS satellite, extending the scope of the said system, serving also users of the Global Monitoring for Environment and Security (GMES) program, like aircrafts, unmanned aerial vehicles and land / marine vehicles. Therefore, the so defined Enhanced Data Relay Satellite Mission architecture would closely relate to a dual (or multi) use mission, mainly used for:

– ordering and collecting data to/from Earth Observation (EO) satellites (including GMES, SPOT, ENVISAT and COSMO-SkyMed), UAV's, launchers, aircraft, maritime and land mobile vehicles;
– support MoD and Space Agencies (ASI, CNES, ESA, NASA, JAXA and others missions), controlling the above platforms (TM-TC&RNG) to update their mission planning in such a quick reaction mode, that can be considered, when needed, real time or nearly-real time response.

The main considered system requirements are as follows:

– multi-purpose mission design, including data relay service, communication to/from mobiles and broadband connectivity, in order to make the security satellite network effective;
– dual-use, for civilian applications and for Italian and European security needs;
– phased mission approach, considered as suitable to achieve capabilities as early as possible for ARTEMIS quick replacement and to extend the network, according to timely needs and gradually extending the capability set for extension to multiple different user classes in scope of the DRS (see DRS roadmap in Figure 15.5);
– adoption of inter-satellite links based on both RF links and optical links;
– secure and anti-interference data links;

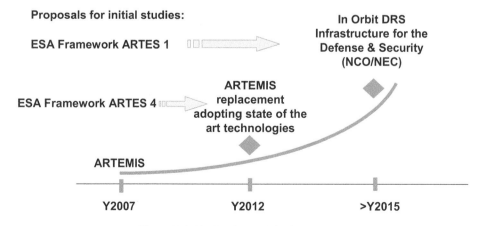

Figure 15.5 The Roadmap of the dual-use DRS

- seamless integration of space-communication networks with terrestrial communication networks;
- feeder link coverage on the European land areas;
- highly flexible and modular payload concept, to be compatible with small GEO platforms but with growth capability for sizing and accommodation on an @bus platform;
- modular DRS satellite conception, peered by the architectural definition of space and ground users of the DRS mission, including LEO terminals, aircraft/UAV terminals, FSS/MSS/BSS terminals;
- interoperability with other DRS existing systems (TDRSS) and backward compatibility with ARTEMIS Users and Ground Segment, the need to fit into the existing regulations and the need to provide secure data links as required by the application;
- compliance with regulatory bodies, and in particular with ESA/NASA/JAXA Spectrum Policy in use, for the Ka, V bands for tracking and data relay service.

As regards the User Requirements, Figure 15.6 provides in detail the near and long term utilisation scenario, for the most promising scientific and institutional missions.

On the basis of the user requirements above mentioned, the following Service Area requirements have been identified:

- on-earth permanent Wide Service Area (WSA): extending the ARTEMIS European FL coverage area, to include the EU25 countries;
- on-earth Steerable spot-beam Service Area (SSA): contacts from/to other countries anywhere on the global coverage seen by the DRS satellite;
- Inter Orbit Link (IOL SZ) Service Zone: optimised on the basis of selected USTs to be served LEO, MEO satellites, space station and automated transfer vehicles, unmanned aerial vehicles, launchers, minimising the Zone of service Exclusion (ZOE);
- Theatre spot beams Service Area (TSA): at least 4 steerable theatre spot beams, anywhere on the global coverage seen by the DRS satellite, with beam diameters in the order of 1000 Km.

IP Protocol awareness should be guaranteed by the use of DVB standards for IP flow transport on the satellite interfaces (both inter-satellite links and feeder links) and the IP over CCSDS services for the higher protocol layers.

The proposed DRS generic target reference configuration is shown in the Figure 15.7. The overall target system will consist of the Space Segment and the Ground Segment. The Space Segment consists of:

- A GEO satellite system, defined in a modular way with spacecraft bus (and resulting embarked payload building blocks) ranging from a small GEO

User	Application		Current Service Provider	Customers Examples
Scientific	Space Observ.	Scientific data TT&C	DSN - TDRSS	FUSE  Gravity Probe B  Hubble Space Telescope 
	Earth Observ.	Scientific data, imaging TT&C	DSN - TDRSS - KSAT (Kongsberg Satellite Services) ARTEMIS	AURA 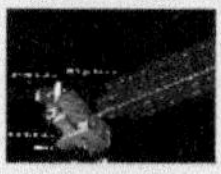 EO-01  ERBS  Landsat 7  ENVISAT  SPOT  COSMO-SkyMed II 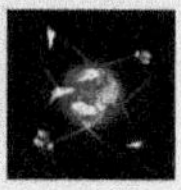 Sentinels 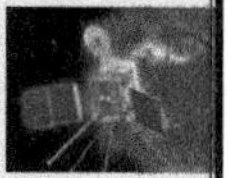
	TT&C and Communications for manned spacevehicle		TDRSS (ARTEMIS)	Automated Transfer Vehicle  Space Shuttle  International Space Station 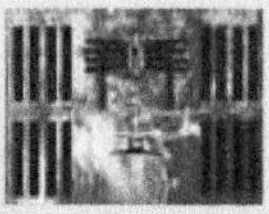
Institutional	Security services Navigation/ Positioning			GALILEO/GALILEO II 

Figure 15.6 Candidate applications and related user requirements

platform, through state of the art platforms and up to the large size platforms (e.g.@bus);

– User Space Terminals, exploiting the DRS system through the Orbit Links for improved visibility and capacity performance and possibly to reduce on-ground operational and cost issues, i.e. by relying on a fully distributed ground segment and avoiding the development and installation (or the rent) of large earth stations.

The Ground Segment consists of:

- User Earth Terminals, constituting the on-ground distributed traffic user installations for traffic and control connectivity with the User Space Terminals;
- DRS Mission and Satellite Control Centre, including the TT&C, Satellite Control Centre and Mission Control Centre elements.

The DRS System will be able to provide the transmission means to support the following required traffic types involving on-ground users and earth stations:

- broadcast traffic from gateway stations on the European coverage (WSA) towards mobile or nomadic stations in theatres of operation;
- bi-directional star traffic between gateways stations in European countries to/from: VSAT terminals on the European (WSA) coverage, mobile or nomadic stations in theatres of operation (TSA) and spread mobile or nomadic users outside theatres (LSA);
- point-to-point two-way or point to multipoint one-way traffics among mobile or nomadic users inside a theatre or in different theatres (TSA) and among terminals in European coverage.

The Figure 15.8 outlines the main traffic types and identifies a preliminary distinction between vital and desirable flows, highlighting the role of WSA and TSA coverage for DRS support purposes.

A possible DRS mission architecture (depicted in Figure 15.9) is composed of:

- Master GEO DRS Satellite, constituting the central service infrastructure and traffic capacity provider;
- GEO piggybacks and opportunity payloads, providing the required extension of the Service Zone, flexibility in terms of traffic evolution demands

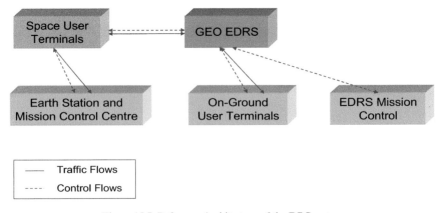

Figure 15.7 Reference Architecture of the DRS system

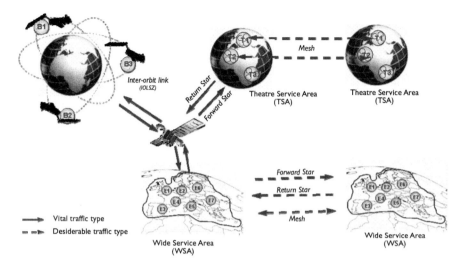

Figure 15.8 Reference EDRS supported traffic types for ground users

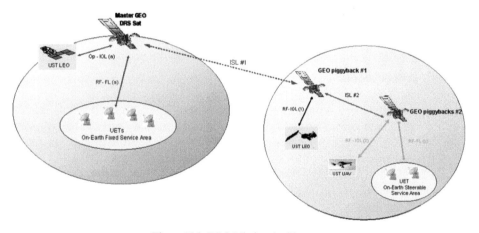

Figure 15.9 DRS Mission Architecture

from on-space and on-ground users, during the lifetime of the central satellite.

Promising Frequency Bands, assigned for DRS service, are:

- IOL: Ka band (23/26 GHz), V Band, Optical;
- Feeder Link: Ka band (20/30 GHz) and Q band. At DRS satellite level, the following configuration of 6 subsystems represents the reference solution (see Figure 15.10):
- Steerable antennas (4 for Theatres and 2 for IOL Space links);
- Optical terminal;

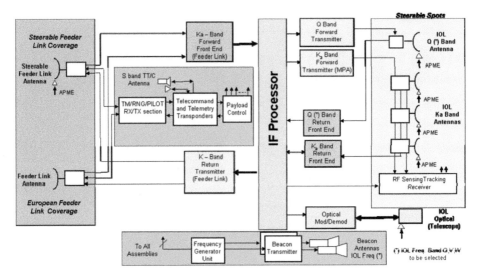

Figure 15.10 Dual Use DRS satellite Payload Conceptual Diagram

- Feeder Link antenna (Euro-Feeder Link and Steerable Beam);
- TX segment (K and V band);
- RX segment (Ka, V, Q);
- IF and Common Unit segment.

As final remark, it is pointed out that the ESA-ARTEMIS satellite will operate still for few years (until Y2012 tbc), with degraded orbital performance (North-South: $+/-4°$ inclination): urgent decisions, about DRS future need to be taken, to guarantee DRS service for the existing operational LEO Users (ENVISAT, ATV, SPOT), as well as for the new candidate missions and launchers. In this respect the agencies (ASI, ESA, ...) can play a fundamental role to support the development of the DRS system and to guarantee the continuity of the involvement of Italian industries/universities/research organisation in the DRS systems context.

15.5 Concluding remarks

The above mentioned two initiatives can provide huge benefits to improve and augment the ISINS infrastructure, especially in the optic of implementing "System of Systems" supporting NCO's and C4I systems. The information, according to the new evolution/doctrine to support NCO, can be delivered across all levels of decision and command, interconnecting multitudes of sensing systems, to provide the "Shared Situational Awareness" that today represents one of the more demanding functions for the NCO. The two presented space components certainly can offer redundancy and robustness against unavailability of the terrestrial communication systems, with coverage areas, capacity, timeliness and performance that the new intelligence sensing and recognition systems need, especially in high mobility conditions. A number of actions are on going, by the Italian Industry, to proceed on the above two initiatives, and to maintain leadership positions and primary roles. In particular the Italian Industry is ready to closely work with Italian MoD and ASI to consolidate the decisions about the above two initiatives and to proceed with concrete implementation plans.

Chapter 16

Future Trends of an Integrated Aero-Space System for Autonomous Unmanned Vehicle Missions

S. Vetrella, L. Verde, U. Ciniglio, F. Corraro

C.I.R.A. (*Centro Italiano Ricerche Aerospaziali*)

The national and international Airspace System (AS) is going to be shared, more and more, by all users, manned and unmanned, to support civil and commercial applications, homeland security and national defence. Unmanned Aerial Vehicles (UAVs) must be integrated into the existing AS infrastructure while enabling safe, efficient and effective operations. In this framework, the National Italian Aerospace Program includes the design and development of unmanned subsonic and hypersonic flying laboratories (CIRA UAV and USV projects), which allow CIRA to test those enabling technologies and methodologies required by the Italian industry to produce advanced future UAVs In this chapter the CIRA vision of a new integrated aero-space system is presented and some key technological challenges are shortly described.

16.1 Introduction

The full integration of advanced Unmanned Aerial Vehicles (UAVs) within the national and international Airspace System (AS) requires a vehicle flying autonomously in an "Unstructured Environment" [1] [2]. "Automation" refers to the absence of human intervention, and "unstructured environment" is associated with uncertainty both in the outside world (meteorological conditions, air traffic, fixed and moving obstacles) and in the vehicle subsystems (failures). The desired level of autonomy can be achieved only through a system able to simulate, even with a better reliability and efficiency, the classic human process: "See-Understand-Decide-Act".

"See-Understand" refers to the capability to have an adequate awareness of the outside-world and of the vehicle behaviour, by acquiring real-time data through advanced sensing systems and processing the information provided to understand "what is going-on". "Decide and Act" requires, on the basis of the "see-understand" process, advanced algorithms to decide "what to do" and to fly the vehicle using software and hardware actuators. For example, the on-board autonomous flight mission management system should be able to modify the flight mission plan on the basis of various evolving conditions, to autonomously land and take-off on a

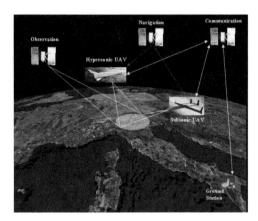

Figure 16.1 An Integrated Aero-Space Scenario

selected runway, to generate and execute safe avoidance manoeuvres with respect to flying and fixed obstacle, etc.

The existing and future navigation and communication satellites are a significant element of the integrated aerospace system, being able to guarantee accurate and reliable navigation data and high data rate communication channels to interact with other manned and unmanned players.

The space segment includes also remote sensing satellites, which are mostly limited to give the mission manager the capability to program and update the mission plan, without a direct satellite-UAV link, due to the high data rate and processing speed required for real-time night and day all-weather vision at high geometric and radiometric resolution (see Figure 16.1).

This chapter presents an innovative concept of a *Global Imaging Illumination Space System* (GI^{2}S^2) to improve the autonomous flying capability of the UAV by means of an on-board *Forward-looking Receiving-only Synthetic Aperture Radar* (FLSAR) and gives a short overview of the technological challenges addressed at CIRA in the field of on board autonomous decision making process, with specific reference to adaptive end-to-end autonomous navigation, autonomous landing, autonomous collision detection & avoidance and adaptive fault tolerant control.

16.2 The Global Imaging Illumination Space System

During various phases of the mission, the "see-understand" process finalized to autonomous flight and operation requires acquiring and analyzing high geometric and radiometric resolution all-weather day and night images.

These requirements can be satisfied by a space borne side-looking Synthetic Aperture Radar (SAR) or an airborne Forward Looking SAR (FSAR). In both cases, real-time forward-looking is necessary to allow the vehicle to see the "environment" along the flight path. A space borne SAR could offer the UAV forward-looking of the area along the flight path by using a side-looking SAR, acquiring

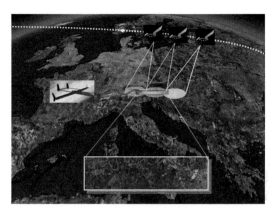

Figure 16.2 The Forward SAR Concept

the raw data, processing and transmitting them to the flying vehicle through a high data rate communication channel.

Global coverage, swath width, geometric and radiometric resolution, time delay in data processing and transmission, additional hardware complexity on board the UAV, limit the potentiality of a cooperative approach between the satellite and the UAV.

On the other hand, conventional airborne F-SAR can not achieve a high azimuth resolution due to the small Doppler frequency gradient and the azimuth ambiguity of terrain targets situated symmetrically about the flight path of the UAV.

Various authors have proposed to improve the geometric resolution of the image by using multiple receiving antennas and processing algorithms, increasing the complexity of the system [12] [13].

This chapter proposes a new approach, based on a Global Imaging Illumination Space System (GI^2S^2), that is a constellation of satellites illuminating the flight path of UAVs by means of a cooperative multiple-beam technique, leaving to a Forward-looking Receiving-only SAR (FRSAR) the acquisition of the raw data and the required processing on board the UAV (see Figure 16.2).

Several advantages can be identified with respect to the above mentioned space borne and airborne techniques, such as:

- Range and azimuth resolution of the FRSAR depends on the frequency, chirp and pulse repetition frequency (PRF) of the satellite transmitter: this allows to overcome the small gradient of the Doppler frequency of a conventional FSAR (i.e. to improve the FRSAR's azimuth resolution) and to increase the pulse duration of the space borne radar (i.e. to improve the FRSAR's range resolution), due to the extra-time offered by the absence of signal reception on board the satellite.
- The Signal to Noise ratio of the FRSAR benefits of a shorter distance between transmitter, target and receiver, with respect to a spaceborne SAR,

and of the optimal bistatic observation angle which can be achieved by taking into account the scattering pattern of the scene.

–   Forward-looking real-time high resolution interferometry can be achieved, offering the UAV a three dimensional vision, which allows to use a further dimension in analysing the environment and in discriminating among various objects and between a flying object and the background clutter.

–   The UAV hardware complexity and mass is reduced, with respect to an airborne FSAR which requires the transmitting radar chain and multiple antennas.

–   The UAV is less detectable and more jamming free due to the receiving-only chain and the variable angle of observation.

16.3 On-Board See & Understand Subsystem

In view of a dedicated experimental mission, CIRA is going to test the FRSAR technology by using the COSMO-SkyMed constellation, with the objective to improve the existing performance of its flying UAV, as far as landing, target identification and autonomous fixed and moving obstacle detection capabilities are concerned.

The main challenge is to achieve high resolution three dimensional data acquisition and real-time on-board processing, to observe the ground, to identify targets and to choose, in emergency, the landing area and detect obstacles. The FRSAR will be integrated in a multi-sensor suite consisting of a low resolution pulsed Ka-band radar, two visible (panchromatic and colour) and two infrared (IR) videocameras, two computers, one dedicated to sensor fusion algorithms, the other devoted to image processing [4]. The system guarantees 10 Hz real-time, all-weather, all-time operations with a field of view of 120° azimuth, 18° elevation and 6 Nm range.

The system has already been validated using numerical and real-time HW-in-the-loop test, while several flight tests are planned in 2008.

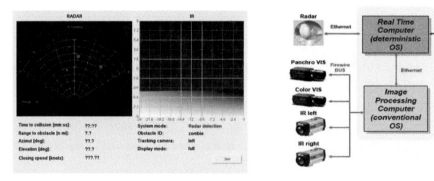

Figure 16.3 Sensor Fusion for Obstacle Detection

16.4 On Board Decide & Act Subsystem

The need to increase safety and capacity of the commercial air traffic system requires the development of more sophisticated and flexible guidance techniques than those presently in use. The innovation expected for the near future air traffic control system necessarily proceeds through an improvement of the on-board autonomous guidance systems, capable of managing various operational events independently from the human intervention.

The autonomous guidance system is even more essential to enable the expected development of the unmanned vehicles, from platforms used exclusively in the military sector, to systems completely integrated in the commercial air traffic system and capable of performing even more complex operations. In the specific case of unmanned vehicles, a further strategic objective is to increase the reliability level of the platform to make it compatible with the requirements applied to manned flights [11], without however increasing costs of development and maintenance of the platform itself. Thus, the research effort towards the above cited challenges is mainly focused on three key technology areas related to: autonomous end-to-end guidance and navigation, autonomous collision avoidance, adaptive fault tolerant guidance, navigation and control.

16.5 Autonomous End-To-End Guidance & Navigation

Autonomous End-To-End Guidance & Navigation refers to the capability of the UAV to autonomously execute missions: to take-off and land over selected areas, to compute and track optimal trajectories, to reach the target area, to face unexpected situations, to execute the mission goal and to come back to the landing site.

Current technologies require a heavy on ground operator support and pre-mission planning, in order to program the UAV system to perform the mission.

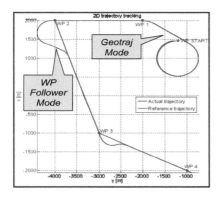 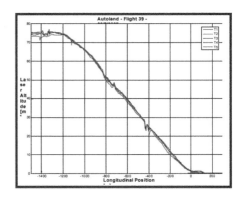

Figure 16. 4 On-Line trajectory Generation and Adaptive Landing Algorithms at CIRA — In flight Validation

In parallel with the design and development of its High Altitude Long Endurance UAV, CIRA is developing a challenging research project (TECVOL) to test and in flight validate a software/hardware prototype by using a VLA (Very Light Aircraft). Several flights have been already performed to test autonomous landing and end-to-end mid air guidance and navigation, while the full autonomous mission is planned to be flown in 2008. The following functionalities have been already tested during various flights:

– Autonomous generation and tracking of an optimal trajectory (minimum time) with assigned final position and velocity by taking into account aircraft manoeuvre dynamic constraints (i.e. stall, load factors and envelope limitations) [6].
– On-line adaptive generation and tracking of approach and landing trajectory on a runway equipped only with a DGPS ground station [3].
– Autonomous Navigation integrating proprietary sensor fusion algorithms with a multisensor AHRS-DGPS-ADS and Laser Altimeter architecture [3].

Many technologies are under development or will be demonstrated during 2008, such as adaptive 4D End-To-End Autonomous Navigation (on-line way-point generation with time constraints adapted to changes of traffic and environmental conditions), GPS-EGNOS based Autonomous take-off and Landing (does not require instrumented runways and is compatible with the future GALILEO system).

16.6 Autonomous Collision Avoidance

Aircraft mid-air collision is still an unresolved problem, as available mishap data show [8]. This situation is going to become even worse with the increasing emerging traffic of small business aircraft, General Aviation and VLA operating from and to secondary airports. Taking into account also the further burden due to operational UAVs, a significant increase of the airspace capacity becomes a fundamental condition for the future development of flight, and, consequently, a robust autonomous collision avoidance system must be designed, developed and put in place [9], which shall guarantee reliability and low impact on traffic rerouting (minimization of the deviations from the flight plan).

It is worthwhile noticing that most of the existing algorithms are not suitable for real-time applications, because of the nondeterministic computational time needed for taking a decision. For example, algorithms have been proposed in which conflict resolution is formulated as a constrained optimization problem, with the aim of finding trajectories which minimize a proper cost function. The main drawback is that computation time is not predictable and the convergence to a solution is not ensured in a finite and deterministic time interval, as required by a real-time control system.

In this field, CIRA has developed a proprietary algorithm [5] allowing on-board autonomous collision detection and avoidance manoeuvring, given that oncoming

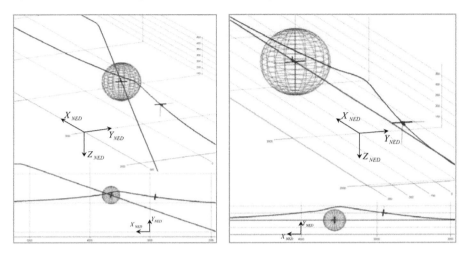

Figure 16.5 Collision Avoidance Manoeuvre generated by the CIRA algorithm

traffic is tracked and identified at adequate data rate. The collision detection & avoidance algorithm does not require cooperation with the oncoming traffic (no information on present and future plans) and only uses information provided by the on-board obstacle detection system. Also, key features of the algorithm are a precise collision detection and minimization of deviation from the nominal trajectory by performing a full 3D manoeuvre.

A preliminary validation of such algorithm has been performed at DLR Air Traffic Simulator of Frankfurt Airport, financed by the EC (IFATS project). Flight tests are currently planned in 2008 with two Very Light Aircrafts, while it is still in development a robust avoidance algorithm to be used with multiple intruders, which is suitable for its use in the Civil Airspace, as the avoidance manoeuvre is computed in order to minimize deviation from the reference trajectory (minimum impact on traffic re-routing).

16.7 Adaptive Fault Tolerant GN&C

How to achieve high performance and reliability against various unforeseen events, failures, uncertainties and other changes in plant dynamics has been a very challenging issue for control system design in recent years [2]. Reconfigurable flight control has the challenging objective to guarantee greater survivability, owing to its capability of automatically redesign control laws and adapt to failures and damages.

A number of different approaches for control reconfiguration upon failures have been proposed and developed in the past years [10]. One of the most promising approaches is based on the use of adaptive model following control techniques, providing that the methods are suitably adapted to failure reconfiguration purposes. This approach is typically classified into two methods: *Indirect Adaptive Control* and

Direct Adaptive Control. The former is based on a parameter identification process, whereas the latter estimates directly the optimal controller gains without the need of exact information about system behaviour.

CIRA works on this topic have been focused on both reconfiguration capabilities for sensor failures using a virtual sensor concept (i.e. the measurement coming from the faulted sensor are substituted by an algorithm that uses measurements coming from the remaining operative sensors) and on adapting the flight guidance and control laws upon detecting actuation or structural failures. In this last case, the algorithm integrates a direct adaptive control technique and a dynamic control allocation algorithm to maintain adequate levels of flight safety against control surfaces failures by means of an optimal use of the healthy ones.

Reconfiguration algorithm to sensor failures have been developed and fully implemented for the next DTFT2 mission of CIRA-USV program (planned in 2008), while an aircraft Stability and Control Augmentation system with fault tolerant capability has been developed within a project financed by Italian DoD, and validated via numerical simulation on a tailless airframe advanced configuration.

CIRA running activities relates mainly to the integration of the above mentioned technologies in one fault tolerant GN&C architecture with emergency replanning and reconfiguration capabilities, able to guarantee at least the safe UAV recovery against failures of critical aircraft subsystems (engines, airframe, navigation sensors, fuel system, etc.).

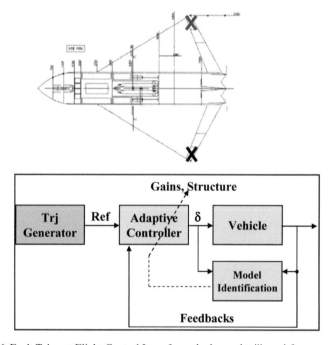

Figure 16.6 Fault Tolerant Flight Control Laws for and advanced tailless airframe configuration

16.8 Conclusions

This chapter presents the CIRA vision and approach toward the design and development of unmanned subsonic and hypersonic Aerial Vehicles, in view of their operational integration within the national and international Airspace System. The "See-Understand-Decide-Act" is the classic human process that CIRA is trying to simulate to give the capability to UAVs to fly autonomously in an "Unstructured Environment".

To this end, the chapter, besides the description of the software/hardware prototypes currently under test by using a Very Light Aircraft (VLA), proposes a Global Imaging Illumination Space System to improve the "see" capability of UAVs. The system takes advantage of a multi-beam space transmitters' constellation to give the UAV a night-and-day all-weather vision at high geometric and radiometric resolution using a Forward-looking Receiving-only Synthetic Aperture Radar.

This new concept will be tested in a near future by "catching" the X-band chirp transmitted by the COSMO-SkyMed constellation through a receiving-only radar on-board the VLA CIRA UAV.

Further research activities are anyway requested to develop new processing algorithms and to integrate 2-D and 3-D SAR images into the existing multisensor suite.

The chapter finally shows how technology research and flight demonstration activities performed in the framework of the National Italian AeroSpace Research program will enable the national industry to play an even more important role in the envisaged future integrated aero-space scenario.

References

[1] Clough, B. T., "Unmanned Aerial Vehicles: Autonomous Control Challenges, a Researcher Perspective," 2nd AIAA "Unmanned Unlimited" Systems, Technologies & Operations, September 2003.
[2] Patcher, M., Chandler, P.R., "Challenges of Autonomous Control," *IEEE Control System Journal*, August 1998.
[3] Ciniglio, U., Corraro, F., De Lellis E., et alii, "UAV Free Path Safe DGPS/AHRS Approach and Landing System with Dynamic and Performance Constraints," *UAV SYSTEMS 2007 — International Technical Conference & Exibition*, Paris, France, 12–14 June 2007.
[4] Accardo, D., Moccia, A., Ciniglio, U., Corraro F., et alii, "Multisensor Based Fully Autonomous Non-Cooperative Collision Avoidance System for UAVs," *AIAA Infotech@Aerospace Conference 2007*.
[5] Carbone, C., Ciniglio, U., Corraro F., and Luongo, S., "A Novel 3D Geometric Algorithm for Aircraft Autonomous Collision Avoidance," *45th IEEE Conference on Decision and Control*, San Diego, California, December 13–15, 2006.
[6] Ambrosino, G., Ariola, M., Ciniglio, U., Corraro, F., Pironti A., and Virgilio, M., "Algorithms for 3D UAV Path Generation and Tracking," *45th IEEE Conference on Decision and Control*, San Diego, California, December 13–15, 2006.
[7] Carbone, C., Ciniglio, U., Corraro F., and De Lellis, E., "Autonomous Decision-Making Algorithms for UAV Flight Management," FOI WORKSHOP 2006.
[8] FAA Aviation Safety Data, FAA website.

[9]   25 Nations for an Aerospace Breakthrough — "European Civil UAV Roadmap" — Final Report, Thematic Network on European Civil UAV FP5 R&D, 2005.

[10] Zhang, Y., Jiang, J., "Bibliographical Review on Reconfigurable Fault-Tolerant Control Systems", *Proc IFAC-Safe process 2003*, Washington, June 2003.

[11] USA DoD, *Airspace Integration Plan for Unmanned Aviation*, internet white paper http://handle.dtic.mil/100.2/ADA431348.

[12] Dai, S., Liu, M., and Sun, Y., "The New Development of High Resolution Imaging for Forward Looking SAr with Multiple Receiving Antennas," *Proceedings of SPIE*, Vol. 4548, 2001.

[13] Dai, S., Wiesbeck, W., "Imaging Mode of Forward Looking SAR with Two Receiving Antennas," *Proceedings of IGARSS '99*, June 1999, Hamburg, Germany, Vol. 3.

PART 3

Dual Use Applications

Chapter 17
Electronic Defence Dual Use Technologies and Applications

Andrea De Martino, Sergio Attilio Jesi

Elettronica S.p.A.

17.1 Background

Elettronica SpA, founded in 1951, is one of Europe's leading manufacturers of Electronic Defence equipment. The experience in design and production of Electronic Defence systems, gained in more than 50 years, enables the Company to guarantee a reliable, effective and consistent response to the ever changing requirements of modern defence. Elettronica is the key word to all solutions, systems and products that comprise Electronic Protection: from the capability to intercept within law enforcement operations and the surveillance of sensitive sites, to the self-protection of platforms in hostile areas and close e.m. spectrum monitoring.

The Company's products line covers all aspects of electronic warfare, from single stand-alone equipment to complete integrated systems for naval, airborne and ground applications, that are in service with the Armed Forces of 28 nations in 5 continents.

More specifically, Elettronica SpA is specialized in the design, development and manufacture of passive electronic warfare equipment for search, detection, analysis, identification and localization of electro-magnetic equipment emissions; electronic countermeasure equipment; radar warning receivers; integrated systems. Based on proprietary technologies, these systems are the core of self protection suites for high performance aircraft, early warning and maritime patrol aircraft and helicopters, naval applications for major warships. Elettronica SpA is also active in the field of ground-based electronic defence systems, including data collection, surveillance and situation awareness for both fixed and mobile installations, associated to electronic warfare analysis centers.

The technologies, the products, the systems developed are essential components of Information Warfare within a Network Centric Infrastructure devoted to Civil as well as Military Defence.

The Company has a proven record of successful international collaborations with platform manufacturers as well as with other electronic industries on a world-wide basis. These long-standing collaborations have led to the definition and implementation of important systems such as the EW suite for Consortiums and International Programs, such as Typhoon Air Superiority Aircraft (UK, GER,

SPA ITA), Horizon & FREMM Frigates (FRA, ITA), NH90 (FRA, GER, NL. ITA), Tornado Multirole Aircraft (UK, GER, ITA), EH101 (UK, ITA), AMX fighter (BRA, ITA).

Moreover, to meet the new operational requirements of the asymmetric threats in the XXI century, Elettronica SpA is currently investing in IR countermeasures as well in high speed monitoring over the entire band width radar & communications EM spectrum, on ground, on board and, in perspective, on space.

17.2 Operational Requirements and Increasing Importance of Dual Use Technologies

In the new scenarios of asymmetric conflicts, the differences between civil and military requirements are steadily decreasing. Today, in the United States, "homeland security" is managed with procedures, requirements, equipments, budgets derived by those of the Department of Defence. Also, Europe and its nations, even if with delay and with more limited resources, is handling a similar process.

Prevention, intelligence, sensor networks, data integration, protection and self protection and reduced time of reaction are key elements of a common approach, civil and military, devoted to increasing the security of the citizens, to react to the increased threats that terrorism has dramatically posed.

In this frame, Defence and Civil Defence are getting closer, at least within western countries and in out of area operations. Network Centric Operations concepts, tested and implemented in the last decades in the Gulf Wars, are utilized also to secure daily used networks such as Transport Networks, IT and Telecommunication networks, Energy and Electricity networks, Financial Transaction networks. It is clear that the vulnerability of these networks vs unpredictable situations, as well, vs hostile actions, is high, as these networks have been always managed with limited "firewalls" to react to these new threats and being such threats very difficult to be detected in advance and with a sufficient time to allow an effective response. Discrimination of false alarms, precise localization of the threat, as well of its nature, are key element to counter react, to decrease casualties. Crisis management and dedicated "high performance" solutions can be found and optimized by an integrated utilization of civil, military and dual technologies. Nonetheless, the challenge is relevant: to re-engineer existing networks, applications, systems within this new paradigm is complex, expensive. To achieve not negotiable results, long term commitments and innovation are required. The risk to implement solutions that save costs but, at the end, are insufficient to defeat terrorism and its way to exercise its threats is relevant.

17.3 Dual-Use Technologies: A Common Definition to Avoid Misinterpretation

Dual-Use Technologies can really help in developing new products and applications allowing the necessary response to "post 11th of September 2001" scenarios.

Unfortunately, wording to define them is very broad in technical literature, leading to misinterpretation, worse, postponing investments and related innovation processes.

In Elettronica SpA's view, dual-use concepts are applicable only to common, civil and military, operational requirements, being satisfied by interchangeable products.

In this frame, generally speaking a fighter aircraft and a commercial aircraft do not utilize dual-use technologies. But, a passive emitter tracking system that allows the surveillance of hostile aircraft as well the surveillance of small commercial aircraft flying in forbidden routes or at forbidden elevations, utilize dual-use technologies.

With an agreed and not questionable definition, the possibility "to do the right things in the necessary timeframe" will become higher and beneficial to the security of million of persons.

17.4 Comprehensive Overview of Electronic Defence Dual-Use Technologies and Applications

Electronic Defence (ED) is a vast military technological domain that comprises:

- Passive sensors systems for interception/analysis of Radar /Comm signals in order to provide warning about threats to the defended platforms, surveillance about the emitters present in the environment and intelligence operations (Data collection) on specific emitters.
- Active systems for self and mutual protection of platforms against Radar /Comm threats and jamming to Radio Navigation systems (GPS, GALILEO, VOR, DME, TACAN).

The use of civilian technologies (COTS: Commercial Off The Shelf) provided in the recent past and still provide nowadays ED with many benefits in terms of cost-effective performances. In Figure 17.1 and 17.2 the technologies employed in the various ED functions are reported, respectively for the ED Passive Sensors and the ED Active Systems . COTS devices such as Fast Digital Signal Processor (DSP) and Field Programmable Gate Arrays (FPGA) are currently employed for fast digital filtering, demodulation and decoding of signals. High Sampling Rate Analog-to-Digital Converters (ADC) are currently employed for wideband multi-carrier applications and direct IF conversion in the ED Receivers. Figure 3 shows Elettronica's latest Dual-Channel Wide Band Digital Receiver which exploits a very high sampling rate (in excess of 1 GHz) and a multibit ADC, together with a number of powerful FPGAs, to perform the above-mentioned functions and the differential techniques capable of providing accurate signal Direction of Arrival. COTS techniques/technologies such as Artificial Intelligence (AI), Operative Research (OR), and Information Technology (IT) products such as Microprocessors (Power PC/Pentium), Operating Systems (VxWorks), Programming Languages (C/C++)

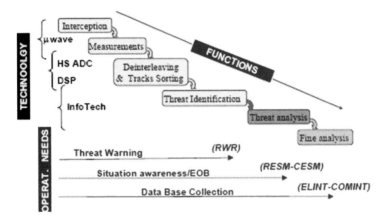

Figure 17.1 ED Passive Subsystems Technologies and Applications

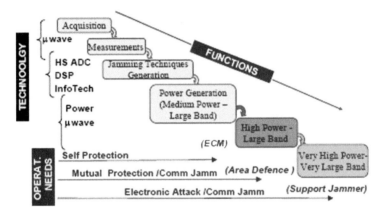

Figure 17.2 ED Active Subsystems Technologies and Applications

and Man Machine Interface (MMI, PC/Windows based), are currently used to perform the following ED Functions/Techniques:

- Emitter De-interleaving in complex e.m. environment, which is exploiting **Artificial Intelligence** and Neural Network based Algorithms to perform the time domain analysis of the emitter pulse sequence (assessment of the PRI law and values) normally affected by missing pulses due to interferences and time overlapped signals. In Figure 17.4 the performance of a Neural Network based Algorithm is reported. It could be noted the improvement provided by such Algorithm in terms of increase rate of missing pulses still allowing the identification of the PRI law.
- Shared Resources Operation of advanced ED system architectures, such as the one reported in Figure 17.5, is managed and optimized through **Operative Research** and **Gaming Theory** applications.

Figure 17.3 ELETTRONICA's High Sampling Rate Wide Band Dual Digital Receiver

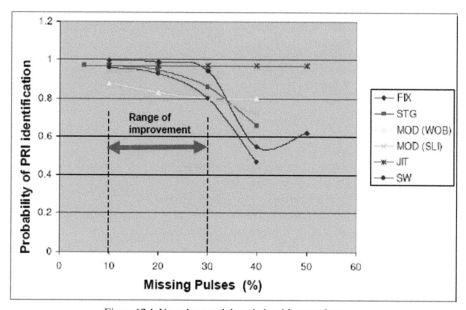

Figure 17.4 Neural network based algorithms performance

- Optimised Interaction with the System by the Operator through the use of **Microsoft Windows**® **OS** based friendly presentation and control. In Figure 17.6 a photograph of the Analysis Mode of an Electronic Support Measures (ESM) Surveillance System is reported, showing how the System Operator can fully master the system with minimal effort.

■ Very large band solid state RX/TX modules and wide band high speed DSP enable the implementation of advanced **ED Shared Apertures Systems integrating passive and active RF functions (ESM+ECM+RADAR +COMM)**

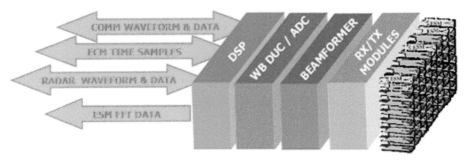

Figure 17.5 Shared Resources Operation of Next Integrated ED System Architectures Generation

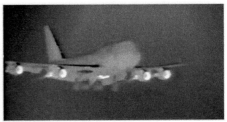

B 747 IR Signature at Take-off   B747 IR Signature at Landing

Figure 17.6 Photograph of the Analysis Mode of an Electronic Support Measures (ESM) Surveillance System

– **Passive Location** of non emitting targets which exploits both commercial transmitters (broadcast, mobile networks, . . .) as well as the above-mentioned COTS devices.
– Multiplatform Tactical Operations (Multiplatform Interoperability, Improved emitter localization using multiple platforms, Cooperative jamming) are performed by ED systems networks which apply the civilian **MANET** (Mobile Ad hoc NETwork) technology which exploits:
  • Rapid deployment of *autonomous* mobile users
  • Communication over wireless radio links
  • Decentralized structure
  • Dynamic topology
  • Stand-alone or connected to larger network

The ED techniques/technologies can in turn be exploited for civilian applications such as:

– **Electronic Protection (EP)** of civilian receivers/sites such as GALILEO (GNSS) receiver on board A/C, helos, High Velocity Trains or Ground-

■The operator's interaction friendly presentation and control , allows full mastering of the system with minimal effort and ensures grow up capability

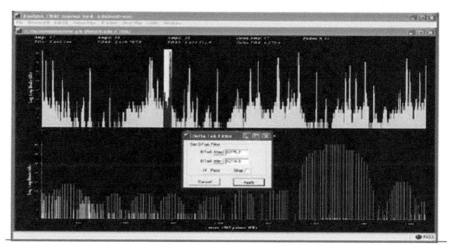

Figure 17.7 User Friendly Man Machine interface based on Microsoft Windows® OS

to-Satellite Communication Stations against jamming signals or non intentional interference. The EP is based on the principle to provide an adaptive nulling antenna system to the GNSS receiver, which is composed of:

- A Conformal Antenna system, properly designed for each platform
- A Common Core comprising:
  - Front-End (direct digital conversion)
  - A/D Conversion e Fast Signal Processing
  - RF Processing.

The schematic block diagram of such EP is depicted in Figure 17.8.

- **Directional Infra Red Countermeasures (DIRCM) System** can be used to protect civilian A/Cs from IR homing ("heat seeking") missiles. This application is based on an Advanced Laser technology (**Fiber Laser**), which provides a powerful laser beam pointing at the missile seeker's head so as to disrupt its homing on the A/C. The system is designed to provide a lightweight and compact solution for the protection of mission-vulnerable aircraft from common battlefield threats. The system is suitable for use in **wide body A/C** such as the B747 (whose IR signature at take off and landing are reported below) **as well as in helo o smaller jet aircraft** for civilian transport. The schematic block diagram of the ELT/572 DIRCM System is reported in Figure 17.9.
- **Passive Location of non-cooperative emitters**, based on the passive multilateration concept, can be used as an auxiliary surveillance system for navigation assistance in emergency situations (ATCR failure). The concept of "**passive multilateration**" concerns with the ability to provide 2D or 3D location of

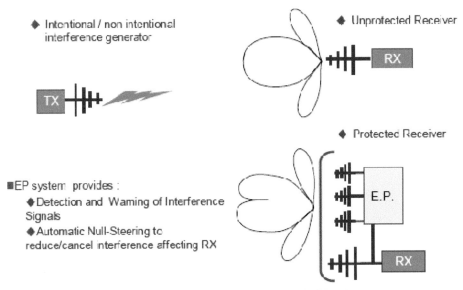

Figure 17.8 Electronic Protection (EP) of GNSS RXs

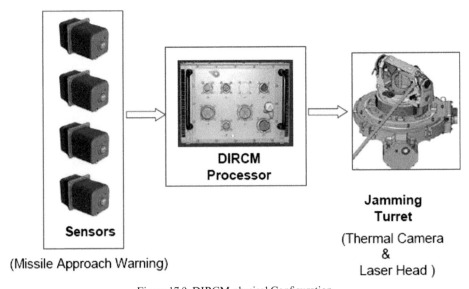

Figure 17.9 DIRCM physical Configuration

steady and moving emitters via measurements of the Time Differences Of Arrival (TDOAs) between two remote ESM/ELINT stations. It has been recognized that "passive multilateration" can bring an **effective contribution to Air Defence Systems**. Indeed passive surveillance featuring the following advantages with respect to radar active surveillance:

■ Modern fast techniques for emitter ranging and accurate location are based on differential measurements of emitter signal phase and time of arrival :

♦ A "twin RX aperture" antenna system, based on two antennas located in different positions, is employed to measure different parameters of the received signal through a dual DRx

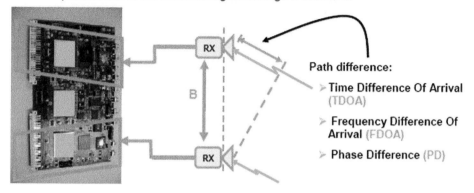

Figure 17.10 Dual DRx Architecture to provide multilateration

- Covert operation
- ELINT identification
- Resistant to countermeasures
- Cost-effective acquisition & lifecycle
- Long range of detection (radio horizon is main limitation) of both land and surface targets
- Excellent tracking accuracy.

Past implementation of military passive multilateration systems has been focused on detection and processing mainly of pulse signals (SSR : Modes 1,2,3/A, 4*,5*, S). Modern DRXs and related advanced high speed digital processing techniques have been developed for military applications requiring location of non-pulse waveform emitters (noise jammer, data link, communications link). Such type of modern military system is a major candidate to the civilian application of multilateration in that:

- All RF sources of the civilian A/C can be used, in principle
- Maximum flexibility is ensured by the DRX approach for the system architecture in view of a cost-effective solution

The Dual-Channel Wide Band DRx on which the multilateration systems are based is reported in Figure 17.10.

As a conclusion, it can be said that nowadays most of the novel technologies are Dual-Use in that they can be beneficially employed in both Civilian and Military Systems.

Chapter 18

Remote Sensing for Civil and Security Applications

Volker Liebig, Alexander Soucek

European Space Agency (ESA)

18.1 Introduction: The Space Perspective

"Fear of Cold War in the Arctic"; "Russians plant Flag on the North Pole"; "New Expedition to substantiate US claims for the Arctic" — three newspaper headlines published in European journals in August 2007. The increasingly political activism in high northern latitudes might have had a very natural reason: climate change. Only a month later, the European Space Agency (ESA) published an article which made it to a front page cover story around the globe — ESA's flagship mission Envisat witnessed the lowest Arctic ice coverage in history. The Northwest Passage, geographical shortcut between Asia and Europe and epitome of impassability, was open.

This episode reveals that climate change is no longer an abstract threat, but a reality. It reveals that satellites, in the quest for understanding the Earth systems, have unique capabilities and therefore a high value. And it reveals that remote sensing, even if undertaken primarily for scientific, i.e. civil reasons, have inherent implications for security and geopolitics.

The strength of a remote sensing satellite is its perspective. The environment of Outer Space allows for an encompassing view along the ground track of the satellite, operating with different instruments and modes and in various wavelengths, orbiting the Earth in some 90 minutes[1], repeating passages over again — for years. ESA's ERS-2 satellite has been launched in 1995 and is still operating. Less physical than legal is another detail: Outer Space is not under state sovereignty, nor subject to national appropriation, and free to use and explore.

Earth observation satellite systems and data are taken for granted today. Newspapers publish "satellite images of the week"; the internet offers satellite-based maps or geo-search engines as freeware. The development leading to this conception has

---

[1] The exact time needed for one rotation around the Earth depends on the orbit properties — especially the altitude — of each satellite.

started decades ago. Meteorology has been the first organised operational community to make use of satellites (notwithstanding the military field). Europe launched the first meteorological remote sensing satellite, Meteosat, in 1977. In the same year, ESA started the "Earthnet" programme to provide European access to non-ESA missions — at that time ESA did not have own scientific remote sensing satellites. This situation changed in 1991 with the launch of ERS-1, followed four years later by ERS-2 and then, in 2002, by Envisat. Envisat is the largest and most complex Earth observation satellite ever built, carrying ten instruments which deliver some 280GB of data daily to ground stations all around the world.

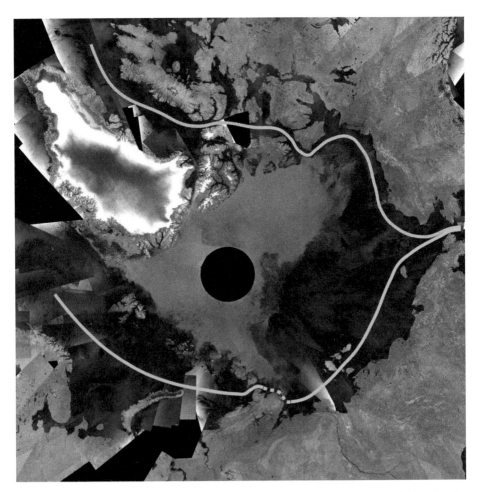

Figure 18. 1 Envisat ASAR mosaic of the Arctic Ocean for early September 2007, clearly showing the most direct route of the Northwest Passage open (orange line) and the Northeast passage only partially blocked (blue line). The dark grey colour represents the ice-free areas, while green represents areas with sea ice. ©ESA

Figure 18.2 Envisat ASAR image of the McClure Strait in the Canadian Arctic Archipelago, acquired on 31 August 2007. The McClure Strait is the most direct route of the Northwest Passage and has been fully open since early August 2007. ©ESA

18.2 Change & Challenge

Oslo, Monday 10 December 2007: The WMO / UNEP "Intergovernmental Panel on Climate Change" (IPCC) and former US vice president Albert Gore Jr. received the Nobel Peace Prize *"for their efforts to build up and disseminate greater knowledge about man-made climate change, and to lay the foundations for the measures that are needed to counteract such change"*[2]. In 2007, the Panels "4th Assessment Report: Climate Change 2007" helped the topic to gain unprecedented media attention through alerting yet substantiated prognoses: a mean global temperature increase between 2.4 and 6.4 degrees until 2100, an ice-free Arctic as of the second half of the 21st century, up to 90% permafrost melting until the end of the century, decreasing precipitation in arid areas, significantly intensified storms and surges, and more. Whereas similar prognoses had already been given earlier, the new report — to a considerable extend based on remote sensing data from satellites — introduced new evidence through longer-term data sets and refined models.

From "climate change" to "global change" it is just one word difference, reminding that any sustainable change of the Earth system, disregarding the discussion on root causes, will inevitably have global effects for our societies. Changes of such kind are not any longer an aspect of a far future, but geopolitical reality. Increasing sweet water scarcity has led to the prophecy that the wars of the 21st century will be about water. Changing patterns of resource exploitation and the looming threat of the end of fossil fuel resources impact security considerations, too. Ecological or economic crises (or both) may lead to migration. And even if the idea from "global warming" to "global warring" is to be taken with some precaution, any deterioration of human living conditions will have consequences.

18.3 Benefits from Satellites: Science, Services, Security

In this situation, the observation of Earth through satellites, both for operational services and (long-term) science, plays an increasing role. Scientists use satellites to enhance our understanding of the correlations of climate variables (and, as consequence, of human influences), whereas the same instruments deliver information that can be applied by a great variety of service sectors.

Water

With Radar Altimeter (RA) instruments as flown on ERS and Envisat it is possible to accurately map the sea surface height (SSH). RA measurements have revealed an alarming trend over the last 15 years — an average global sea level rise between 2.64 and 3.29mm/year.[3] The major part of this rise is caused by thermal expansion of the water due to global warming. Therefore another very accurate indicator, the sea

---

[2]The Nobel Foundation, Prize Announcement, press release, 12 October 2007.
[3]Based on measurements from ERS-1, ERS-2, Envisat, Topex-Poseidon (NASA/CNES) and Jason.

surface temperature (SST) measured by the (Advanced) Along Tracking Scanning Radiometers ([A]ATSR) on the ERS and Envisat satellites, is of high importance. The residual trend in SST raise as measured by ESA's satellites since 1991 is of 0.13°K/decade.

Many scientific methods in Earth observation are mature enough today to be transferred to operational applications. Rapid vessel detection (through Radar satellites), oil spill surveillance, wind, wave & current forecasting, fishery aid or maritime border control are upcoming satellite services. The political dimension of maritime security becomes evident thinking of the *transnational character of maritime affairs*[4], for example for Europe with a coastal length of some 117.000km. The EU has thus adopted an "Integrated Maritime Policy" in 2007, and ESA has concluded a cooperation agreement with the European Maritime Safety Agency (EMSA). The more we use our maritime environment, the more important space-based services and science will become.

Fire

The summer of 2007 brought various devastating wildfire events to southern Europe, North America and several other parts of the world. Again, satellite-based instruments proved to be of high value both for detection and extent measurement as well as for post-disaster assessment. In infrared wavelengths, fires become "observable" from Earth orbit, such as through the radiometer sensors of ERS-2 and Envisat. The observation from "the high ground" allows for an overall view of the damage extend and possible evolution — by the way likewise for fires as for flooding.

Ice & Air

Diminishing sea ice in Arctic and Antarctic zones is not only a warning signal for scientists but a threat for shipping. Sea ice and iceberg monitoring from space is therefore of increasing importance. The trend towards an ice-free Arctic ocean will also increase marine traffic (using the Northeast Passage, the merchant route from Europe to Asia shrinks by several thousand kilometres). Of high importance for ice services are satellite-based radar instruments operating at C-band wavelengths, such as ASAR onboard Envisat. Whereas less ice might free the way to new resources and change the strategic situation for countries bordering the Arctic, it is clear that polar eco-systems face fundamental changes. For 12.000 years, since the last ice age, an Antarctic ice shelf, Larson-B, is thought to have remained stable, until the shelf ice (comparable to the size of the US state of Rhode Island) collapsed in March 2002, observed by Envisat. It was a sign of the vulnerability of the Earth's cryosphere and a reminder that the cryosphere plays an important role for our climate.

---

[4]See ESA, EC unveils new EU maritime policy, published under www.esa.int, 12 October 2007.

Figure 18.3 Crude oil from the wrecked Hebei Spirit tanker is seen polluting the sea off South Korea in this image acquired on 11 December 2007 by Envisat's Advanced Synthetic Aperture Radar (ASAR) instrument. ©ESA

Figure 18.4 On 24 August 2007 Envisat captures billowing smoke from fires raging across Greece's southern Peloponnese peninsula, where fires have claimed the lives of at least 60 people since they began. ©ESA

Satellite-derived air quality data reveal another field of application. Various atmospheric constituents like dust, ozone or nitrogen dioxide can be measured and forecasted (within limits) from space.[5] Like the oceans, also the atmosphere is a complex system not fully understood yet. The scientific exploration of wind patterns, ozone layer, aerosols and many other elements leads also to the development of dedicated space missions: ESA's ADM-Aeolus mission shall provide global observations of wind profiles from space to improve, inter alia, the quality of weather forecasts. The EarthCARE mission shall investigate the influence of water vapour, clouds and aerosols on our climate.

Earth

Due to a red-coloured surface component and its streamlined shape, it is nicknamed the "Ferrari" of satellites: GOCE[6]. The ESA Earth Explorer satellite will fly in an altitude of only 250km (hence its aerodynamic properties in the "vacuum" of space) to determine the Earths geoid with an accuracy of 1–2 cm and the gravity-field anomalies with an accuracy of 10^{-5} m/s^2. This data will teach us a lot about the mass distribution inside our planet. But the geoid serves also as reference detecting ocean current systems in altimeter data, to calculate high precision orbits for satellites or to transfer height measurements with reference to sea level into GPS based values and vice versa.

Land monitoring from space is maybe the most evident task of a satellite for a layperson, and the range of applications is large. Remote sensing is a tool for helping the implementation of the EU Common Agricultural Policy by determining land use from very high resolution imagery. Very high resolution is just as well needed in a very different context from agriculture — for military applications. Verification from space is a means to control international disarmament agreements as well as obtain reconnaissance information: evaluation of usability of military entry points or similar features, vehicle detection and movement monitoring, border control, migration monitoring, structure identification. Satellite interferometry can reveal nuclear tests on ground. X-band SAR allows to distinct between tracked and wheeled vehicles. In this context, it becomes obvious that remote sensing technologies have an inherent "dual use character". They can be exploited for civil or scientific purposes as well as for military purposes. This potential makes cutting-edge remote sensing technology, like other space technologies, still a safeguarded good.

Geohazard mitigation is another application. SAR interferometry, combining two or more satellite radar sources, allows very precise measurements of ground movements — surveying (or revealing) high-risk areas such as volcanoes or seismic faults. Hints for earthquakes, landslides or subsidence of buildings as results of terrain motion can often be detected in advance.

---

[5]PROMOTE, a consortium developed in the frame of GMES under ESA lead, has concluded in early 2008 an agreement with the European Environment Agency (EEA) for the provision of air pollution information.

[6]Gravity field and steady-state Ocean Circulation Explorer.

Figure 18.5 Artist's impression of the GOCE (Gravity field and steady-state Ocean Explorer) satellite. GOCE is the first Core Earth Explorer satellite to be developed as part of ESA's Living Planet Programme and is scheduled for launch from the Russian Cosmodrome in Plesetsk at the end of 2007. ©ESA — AOES Medialab

18.4 New Perspectives in Civil and Security Applications: GMES and Beyond

All these benefits of modern Earth observation have been politically recognised with the advent of GMES, the Global Monitoring for Environment and Security initiative of the European Union and ESA. GMES builds on the role of Earth observation (both from space and in-situ) for an improved implementation of European policy areas. ESA is responsible for the implementation of the GMES Space Component and the coordination of contributing non-ESA missions. This includes not only the development, launch and operation of so-called Sentinel missions (responding to European user requirements) as of 2011 onwards, but also the establishment of harmonised data access to other European, national and private satellite missions, as well as the initial development of dedicated service portfolios (ranging from food security to urban mapping).

GMES will provide a sustainable and independent global observation system for Europe. At the same time it will be the European contribution to GEOSS, a Global Earth Observation System of Systems, which is currently organised by GEO, an intergovernmental body of 72 governments and 52 organisations established in the aftermaths of the 2002 World Summit on Sustainable Development.

Remote sensing from space will continue to be used as essential tool in many ways. ESA will continue, in the context of the European Space Policy as well as the international scientific environment, to develop missions and applications, as the maturity of Earth observation has just started.

Chapter 19

Innovative Satellite Applications for Navigation and Earth Observation

Giuseppe Veredice

Telespazio

19.1 Foreword

At the beginning space industry was born to satisfy mostly military and science needs, but after 50 years from launch of first satellite — Russian Sputnik-1 — space services have moved to get a mass-demand: from governmental, military and scientific institutions to individual end-consumers.

After the boom of satellite telecommunications, Satellite services concerning with positioning and Earth observation are becoming more and more required by several demand drivers.

First, the threat coming from international terrorism all over the world, which increases the need of more advanced solutions for homeland security & safety. Secondly, climate change is going to push up interests for precisely and in real time meteorological information, land and ocean monitoring, and then disaster management.

Satellites usages are common in daily life of each one, but satellite is still just an infrastructure, that is huge costly, thus public procurement is still mandatory to satisfy demand and to sustain space industry. Space services sector are in charge to make real the benefits for tax payers common life of such a technology intensive investments. Space services are one of the most significant source for innovation, knowledge transfer and economical development in many aspect of day-by-day human life.

19.2 Large Space Programs: from "Duality" to Convergence

In the past, "Large Programmes" have been with "dual vocation". They have been committed by Military and/or Institutional Entities and they have been also a civil use.

For instance, GPS, which is a military device, built by the American Department of Defence at a cost of $12 billion and intended primarily for military use, and which is made now available to everyone.

Nowadays, it is going to be more and more common to have satellite systems with "dual use". It means that the procurement comes from military and civil Entities and obviously uses are military and civil.

In the first model the right to switch off "ad libitum" the system remains on only commitment side (i.e. military). Thus the civil usages are just consequently provided.

In the second one, the right to switch off and to extend the system is on both sides, military and civil.

The "dual use" programmes are in fact convergent from their origin, with services and applications specialised for military and civilian customers, but utilising the same platform and within the same system. Civilian Institutions (Agencies) and military ones (MoDs) define together system basic requirements, and identify also separate specific uses: dual nature is achieved "a priori". There are many examples like for instance:

- COSMO-SkyMed, the Earth Observation Italian programme, developed by Italian space agency and Italian MoD;
- GALILEO, European project for satellite positioning system, conceived as a civil / commercial system but likely to be mainly devoted to institutional utilisation;
- GMES, European project for monitoring capability and services and MGCP, NATO cartography activity;
- Athena-Fidus for telecommunications, bi-lateral French and Italian project to provide broadband services for institutional and government customers (security and defence). It is being developed by ASI and CNES to satisfy the needs of Italian and French MoDs;
- Also USA's DoD is pursuing Iris, a satellite routing system for military communications with parallel civil utilisation as well.

19.3 The Important Role of Services

In the space sector different actors play with different goals. On public side there are government institutions (Department of internal affairs, Civil homeland safety, etc.), universities with knowledge's profiles, military institutions, national and international Space agencies. On private side there are large and small/medium enterprises and finally end-users.

The public players have the task to envisage, promote, support and manage the large programs that can lead, drive and foster the economic growth and the industrial competitiveness. On their side, the private companies have the burden to realize the large programs with effectiveness and efficiency, returning to the public authorities the best value for the money they spend. They have the capabilities, the technology skills and the market knowledge that allow them to accomplish their task.

It is in this context that appears the importance of applications and the fundamental role of Service Providers (such as Telespazio) in arranging the use of the space systems.

The continuous improvement of technologies is offering now the development of new families of services, highly innovative and competitive, which increase the capability of monitoring and controlling environment, mobility and sensitive infrastructures.

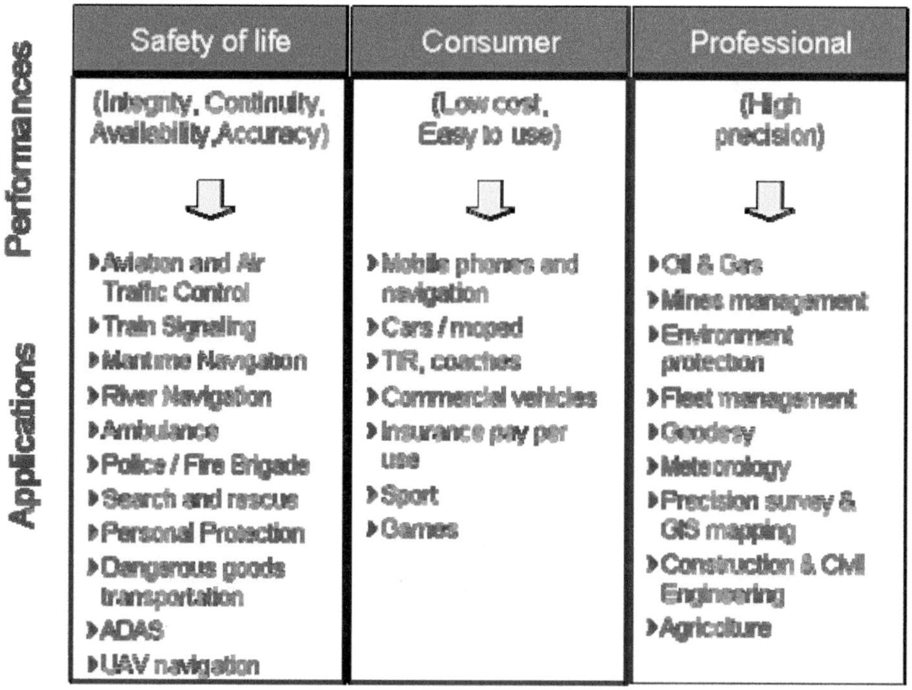

	Safety of life	Consumer	Professional
Performances	(Integrity, Continuity, Availability, Accuracy) ⬇	(Low cost, Easy to use) ⬇	(High precision) ⬇
Applications	▶ Aviation and Air Traffic Control ▶ Train Signaling ▶ Maritime Navigation ▶ River Navigation ▶ Ambulance ▶ Police / Fire Brigade ▶ Search and rescue ▶ Personal Protection ▶ Dangerous goods transportation ▶ ADAS ▶ UAV navigation	▶ Mobile phones and navigation ▶ Cars / moped ▶ TIR, coaches ▶ Commercial vehicles ▶ Insurance pay per use ▶ Sport ▶ Games	▶ Oil & Gas ▶ Mines management ▶ Environment protection ▶ Fleet management ▶ Geodesy ▶ Meteorology ▶ Precision survey & GIS mapping ▶ Construction & Civil Engineering ▶ Agriculture

Figure 19.1 Market segmentation and applications

19.4 GALILEO and the Navigation Services

GALILEO, in spite of the present difficulties, outlines the absolute need for Europe to create Navigation Regulated Services. The main value added, provided by GALILEO, is the certification of Positioning Geo-data. Currently, GPS data is freely available but it does not have any guarantee. Instead, the GALILEO's positioning signal will be sure, guaranteed, certified and controlled in terms of precision — there is a huge discuss about the profile of liability of data and there is not yet a final agreement on it.

New applications will linked with civil homeland security, such as critical safety issues of harbours, airports, railways and rail trucks, and whatever critical and sensitive points of concerning. In addition, we will have services for life rescue in high critical situations and localizations with profiles of positioning and infomobility.

GALILEO Programme has to be seen also as a "political" tool in order to push European growth trend in terms of job creation, economic increase, catalyser for industry at the same time due to SMEs and Large Companies. It will be an excellent vehicle to get a fair synchronization between Large Companies with high financial stability and SMEs with high flexibility in front of new needs.

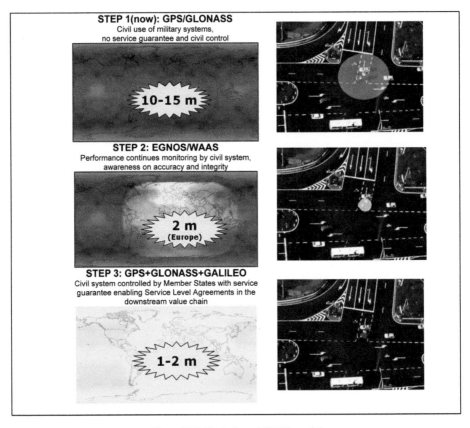

Figure 19.2 Evolution of GNSS model

19.5 Earth Observation Services

Earth Observation satellites are boosting the GMES programme, which is proposing innovative and advanced satellite services integrated with both terrestrial and other space technologies. This initiative is aimed at the establishment by 2008 of a European capacity for the provisioning and use of operational information for monitoring and management of the environment ad for civil security. Italy, France and Germany are contributing to the project with parallel programmes on Earth observation such as, respectively, COSMO-SkyMed, Pleiades and SAR Lupe.

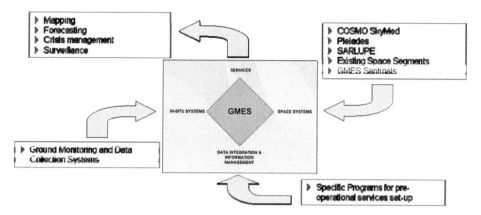

Figure 19.3 GMES programme

Concerning COSMO-SkyMed, it is one of the most innovative programme for Earth observation and represents a flagship of our Country: Italy is gearing up to strengthen its capabilities that will place it on the same footing of the Countries most heavily involved in programmes for protecting the environment.

COSMO-SkyMed opens the era of operational geospatial information applications, no more only for Defence but also for emergency, territory and natural resources management. Thanks to the short revisit time granted by the four constellation satellites and radar technology, it is possible to provide unprecedented operational monitoring services: protection of the territory in such areas as fire, landslides, droughts, floods, pollution, earthquakes and subsidence; management of natural resources in agriculture and forestry; as well as the monitoring of urban sprawl.

Moreover, COSMO-SkyMed peculiarities, such as expandability and multisensoriality, consent the integration and the interoperability with other systems. It will be integrated with two French optic satellites (Helios II) and, thanks to an agreement between the Italian space agency and its Argentine counterpart (Conae), will then work in conjunction with the two Argentine Saocom satellites to create the Siasge programme, dedicated to emergency management.

Figure 19.4 shows the Po river delta seen from COSMO-SkyMed 1, with 5-metre spatial resolution. COSMO-SkyMed's radar sensor allows cultivated land to be clearly shown (varying from light to dark, depending on the vegetation, dampness of the ground, and roughness of the surface) and the identification of areas covered even by a very shallow layer of water (which appear as darker patches in the northern part of the delta). In the sea, the difference between fresh water, rich in sediment washed down by the river, and the waters of the Adriatic criss-crossed by many small craft (whose wakes are clearly visible), is sharply defined.

The future of navigation and earth observation services appears to be virtually unlimited, technological fantasies abound. The space systems will provide a novel, unique, and instantly available address for every square meter on the surface of our planet. Large Institutional Programmes, such as GALILEO, GMES,

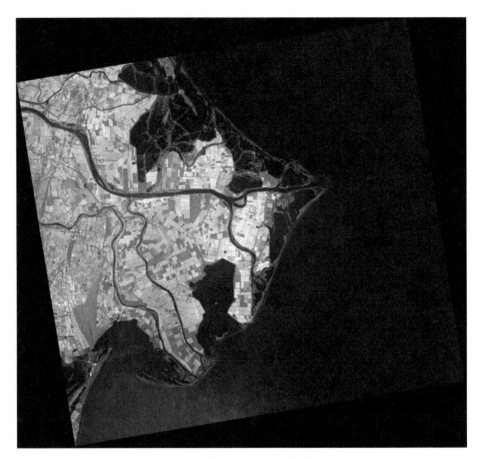

Figure 19.4 Po river delta seen from COSMO-SkyMed 1

COSMO-SkyMed and others previously described, are able to get a high profile of innovation from economic, industrial, technological and security point of view. They will be in certain term quite "revolutionary" as they will improve the degree of quality life of each ones.

Chapter 20

GALILEO: A System with a Dual Use Dimension

S. Greco

Thales Alenia Space

The European Space Policy approach is based on a single "global space policy" merging both the civilian and the military space policies. The European Union contributed to the implementation of this "**dual policy**" by co-funding programs such as EGNOS, GMES and GALILEO.

The GALILEO System (see Figure 20.1) is composed by 3 main segments:

– the Space Segment
– the Ground Mission Segment (GMS)
– the Ground Control Segment (GCS).

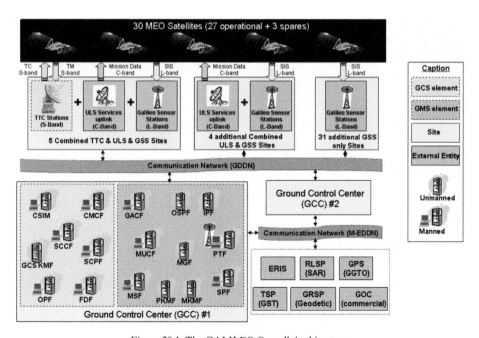

Figure 20.1 The GALILEO Overall Architecture

Figure 20.2 The GALILEO Satellites Constellation

The Space Segment consists of a constellation of 30 satellites (see Figure 20.2) in Medium Earth Orbit (MEO) placed over 3 orbital planes inclined at 54° and at an altitude of around 23,000 km.

The Ground Segments are composed by a worldwide network of 40 Ground Sensor Stations, 9 Up-Link Stations and 5 TT&C Stations controlled and managed by 2 GALILEO Control Centres.

GALILEO is conceived by the EU so as to combine space management and "security and defence" issues. For this reason, GALILEO needs to be highly-protected against already identifiable threats.

Generally, a Dual Use system can be categorised into two types (Type A and B) with respect to the relevant Mission/System Requirements:

- Performance Requirements
- Operational Requirements
- Security Requirements
- Military Needs
- Civil Needs
- ...

From the balancing of the Security related requirements, the system can be of **Type A**, where the Security Requirements are driven by Military Needs and, as a consequence, the Dual System is under Military Control and the products for Civil Applications generally are not guaranteed, or of **Type B**, where the Security Requirements are driven by the Civil Needs and, therefore, the Dual System is under Civil Control and the products for Military Applications are limited by the Security boundary.

In Figure 20.3 the derivation of a Dual System Architecture from the security requirements is depicted.

As it is easy to understand, GALILEO is a Dual System of Type B because the Security Requirements are driven by the User/Civilian needs and the System is and will be under the control of a Civil Authority.

The Security related aspects of GALILEO can be grouped into 3 main areas:

Security of the Infrastructure: intended as the security of the Sites, of the Buildings and, in general, of the Segments that are composing the System;

Security of the Signal-in-Space: meaning the protection of the Signal against Intentional (i.e. Jamming) or non Intentional Interferences and against Spoofing;

Prevention of Misuse of Services: intending the preventive protection and control of the Accesses to the GALILEO Services for hostile use at different levels.

The protection of these Services and of the overall GALILEO System is a fundamental objective and design driver in the framework of the GALILEO development. In particular, GALILEO will embed relevant security countermeasures to provide, among the other services, a Public Regulated Navigation Service (PRS) accessible to Governmental use only with the highest level of protection according to the identified threats and with the performance depicted in Table 20.1.

The final System, as a consequence, will be subjected to relevant security Certification and Accreditation processes.

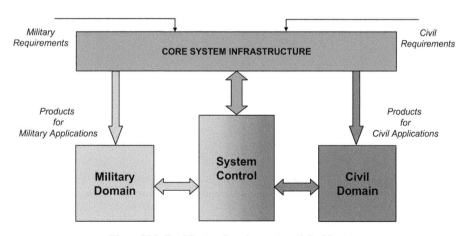

Figure 20.3 Dual System Requirements and Architecture

Table 20.1 Service Performance for PRS with the Satellite Navigation Signals only

	Public Regulated Service		
	Single carrier		Dual
	L1	E6	carrier
Horizontal accuracy (95%)	15m	24m	6.5 m
Vertical accuracy (95%)	35m	35m	12 m
Time Accuracy	Dual carrier		
Timing Lab.ry (Fixed User)	-		
Frequency Accuracy (calibration Laboratory Receiver)	-		
Horizontal Alert Limit	20 m		
Vertical Alert Limit	35 m		
TTA	10 s		
Integrity Risk	2.0×10^{-7} over 150 seconds period		
Mean Availability	99.5%		
Continuity	$1-(8 \times 10^{-6})$ in any 15 s		
Times To Fix			
Cold Start	100 s		
Warm Start	30 s		
Reacquis.	1 s		
Environment	Normal PRS environments		

To provide a Security Accreditation to a system implies to grant a formal state-ment issued by an adequate Authority to confirm that a system (whatever it is) can process, store or forward sensitive or classified information without unacceptable risks, according to specific conditions that are defined in the applicable System-Specific Security Requirements Statement and in the related Security Operating Procedures.

The European Satellite Navigation System named GALILEO is not an excep-tion. It represents, in fact, an information system that provides, according to the GALILEO Security Doctrine, a global service at a strategic level embedding security aspects. This leads to the need of a security accreditation of the GALILEO system in order to state that it is accepted for operational use.

Such Accreditation decision will be made by an adequate authority only after all appropriate security measures have been implemented and a sufficient level of protection of the system resources has been achieved and tested.

This accreditation process is fundamental for GALILEO to implement a "dual-use" dimension.

From the performance point of view it is very important the interoperability and cooperation with the already existing GPS System. The US and the European Union signed on the 27th July 2007 an agreement that will allow their satellite-navigation systems to work together to provide more accurate positioning data. Under this agreement, EU and US satellites will send information on the same radio frequency, enabling receivers to pick up signals from both systems and combine the data. The issue of the interoperability, however, is not only at radio frequency level, but also on the common reliability of the used data. For this aspect, GALILEO will be self-sufficient with the implementation of the Integrity concept, while the GPS is usually complemented by Regional Satellite Augmentation Systems like WAAS in North America and EGNOS in Europe.

The cooperation is then at the level of GALILEO with GPS+EGNOS (or WAAS). In the Figure 20.5 it is possible to notice the improvement on the position

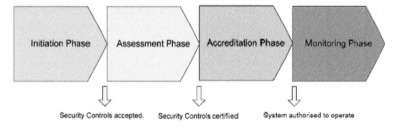

Figure 20.4 GALILEO Security Accreditation Process

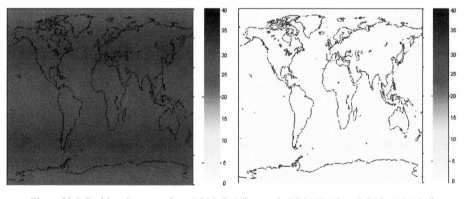

Figure 20.5 Position Accuracy from "GPS-Only" towards "GALILEO and GPS+EGNOS"

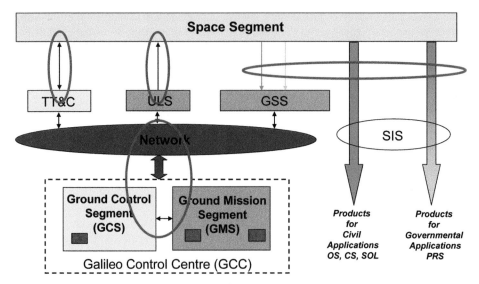

Figure 20.6 GALILEO Security Areas

accuracy passing from GPS-Only towards the GALILEO and GPS+EGNOS combined use.

The improved performance is one of the aspects relevant to the interoperability; the other, but not less important, is the interoperability of the two systems from a "Dual Use" point of view. In order to better exploit the combined "Dual Use" of GPS and GALILEO, it is important to spend a while on analysing their status, policy and evolution.

The GALILEO system has been conceived to support "Dual Use" applications identifying, in its architecture, which are the security areas (see Figure 20.6) to be protected.

Due to the Governmental Use, the PRS Service represents the mean towards the Dual Use of GALILEO and, for this reason, it has been designed having as design driver the Robustness of Performance. Potential "Dual Use" applications that are viable with the use of the PRS are:

– Police
– Customs
– Civil Protection
– Fire Brigade
– Coast Guard
– Law Enforcement
– Regulated/Critical Transport
– Energy and Telecommunications
– Economic, Commercial and Industrial activities that are deemed of National and/or European strategic interest
– ...

It is then shown that GALILEO born to be a Civilian System to be used for Civil Applications and it is going towards a Dual Use dimension with the identification of Governmental applications.

In fact, on the 16th May 2007, the EC Transport Commissioner Jacques Barrot said that GALILEO will be *"civilian controlled* [. . .] *but there will be military users".*

The GPS System followed the opposite way: it has been conceived to be a Military driven system (Type A System) and, in the past 15 years, developed several Civilian Applications culminating with the Civil Requirements considered for the GPS III evolution design.

Also for the GPS, the security issues have been deeply analysed and considered as it is possible to read in the Volpe Report of 2001 [1] where the purpose was to *"assess the vulnerability of the U.S. national transportation infrastructure to degradation or loss of the GPS signal; provide an independent, integrated assessment of impacts to civilian GPS users arising from the degradation or loss of GPS service; provide approaches to mitigate these impacts; and provide the basis for policy decisions on the future navigation system infrastructure".*

The steps of the GPS towards a Dual Use system can be summarized in the following:

1983:   President Reagan offers free civilian access to GPS
1996:   First U.S. GPS Policy. Declared GPS a dual-use system under joint civil/military management
1997:   Congress passes law requiring civil GPS to be provided free of direct user fees
2000:   President Clinton set Selective Availability to Zero
2004:   President Bush issues new U.S. policy on space-based PNT

As the penetration of GPS into civil infrastructure continues unabated, it becomes an increasingly tempting target that could be exploited by malicious persons or countries. If this results in a loss of GPS service, it could, in turn, result in:

–   Transportation service disruption and resulting economic impact
–   Environmental damage
–   Property damage
–   Serious injury or fatality
–   Loss of confidence in a transportation mode
–   Liability to the service provider

In order to manage and drive the policies relevant to GPS, the National Space-Based Positioning, Navigation, and Timing (PNT) Executive Committee was established by the US Presidential directive in 2004 to advise and coordinate federal departments and agencies on matters concerning the Global Positioning System (GPS) and related systems. This Executive Committee produced the policy for the GPS evolution that can be summarized as follow (from Maureen Walker presentation — National Coordination Office for Space-Based PNT):

- No direct user fees for civil GPS services
- Open public signal structure for all civil services
  - Promotes equal access for user equipment manufacture, applications development and value-added services
  - Facilitates open market driven competition
- Encourage use of GPS time, geodesy, and signal standards
- Promote global compatibility and interoperability of future systems with GPS
- Protect the current radionavigation spectrum from disruption and interference
- Recognition of national and international security issues and protect against misuse
- Provide uninterrupted availability of PNT services
- Meet growing demands in national, homeland, economic security, scientific, and commercial uses
- Continue to provide civil PNT services
  - Ensure they exceed, or are at least equivalent to, those of foreign civil space-based PNT services
- U.S. space-based PNT services remain essential components of internationally accepted services

In the similar way, the EU established the Common Foreign and Security Policy (CFSP) defining the following top level objectives:

- to safeguard (...) the independence and integrity of the Union (...)
- to strengthen the security of the Union in all ways

As a consequence, the following drivers have been derived:

- Management of security, linked to the implementation of the GALILEO system, will be exclusively responsibility of EU institutions and EU Member States.
- Specific agreements with non-EU countries wishing to be associated to the project need to be established.
- Policy defined and implemented by EU and Member States through Security documents approved and maintained by GALILEO Security Board.
- Implement security measures to ensure that threats to GALILEO infrastructure and signals, and protection against misuse of signals, are properly addressed.

For what concerns GALILEO, the status, relevant to the Security issues that are the basis toward the Dual Use dimension, is that some issues still remain to be resolved both for GALILEO and more generally for Security and Defence aspects of Space and it has to be considered the application of a Space Security Policy will be more difficult with more Member States.

In conclusions, it is of fundamental importance to exploit key criteria for specific dual applications, such as:

– Priority
– Confidentiality
– Availability
– Readiness
– Security

Furthermore it is not less important to enlarge the perimeter to dual-use applications, through:

– National initiatives for institutional use
– Vulnerability Assessment of various domains by dedicated programs at National/European level.

Bibliography

[1] "*Vulnerability Assessment of the Transportation Infrastructure Relying on the Global Positioning System*", Volpe Report, 29 August 2001.
[2] "*Applied Satellite Navigation Using GPS, GALILEO, and Augmentation Systems*", R. Prasad, M. Ruggieri, Artech House, 2005.

Chapter 21

Convergence of Navigation and Communication Towards 4G

Ramjee Prasad*, Ole Mørk Lauridsen**, Albena Mihovska*

*Aalborg, Denmark; **Terma, Herlev, Denmark

21.1 Introduction

The 21st century has been marked right from its start by the trend for globalization in all spheres of life. The challenges for mobile communication arising from this trend are requiring strong innovation performance and a fast technological pace. The globalization of the economy, the political, cultural and everyday life require also a technological globalization that will ensure that the communication process in all its spheres is including the perspectives of all the main players (i.e., users, vendors, operators, public regulatory bodies, political and governmental bodies). The main elements of this globalization are openness, cooperation, integration, and convergence.

Effective communication between people and the availability of and access to information is increasingly important in nowadays society. This is strongly induced by a growing complexity of the environment in which people operate. Complexity relates to the diffusion of borders, the multitude of impressions and manifestations of products and services, and their — often — virtual character. A greater part of information need is now related to the geographical position of the user.

From a business perspective convergence means new combinations that are required to successfully deploy services that go beyond plain voice and internet access. Such new combinations may lead to "offspring" (new companies), but also to a better understanding, standardization of common interfaces and a joint business model. The aspect of "convergence", joined with the assurance of dependability, will generate the long-awaited breakthrough of information and communication technologies (ICT) in everyday operations. Because of the complex composition of these new services, bringing to the user, this assurance is far from trivial. The vision is that, from a user perspective, one would be able to build up personal networks (PNs) as secure, dependable and trusted entities of the overall infrastructure through a full range of integrated services provided by a range of available technologies. From a technology perspective this means that various devices, networks and underlying radio technologies integrate numerous functionalities, thereby ensuring the personal fit of the communication process, named as personalization. One result of convergence is, for example, the multifunctional handset device that brings to the user: positioning service, music, video, gaming, and streaming functionalities in addition to voice and messaging services. A challenge related to the increasing complexity of the devices is the leveraging of connectivity and dependability with usability, and thereby user-acceptance. By this, MMI becomes a science of its own.

Convergence is a characteristic of the dawning society that will reshape the whole communications, media, consumer electronics and IT industries. Every user must be able to connect everywhere, anytime and with access to adapted and high-quality content and communication services, in a safe and accessible environment.

Convergence is shaped by the choices made by users, scientists, industry and governments, making it difficult to foresee the details of future developments [1]. In other terms, convergence is to telecommunications what globalization is to trade- an issue that will affect everything that governments do [1]. The current trend in mobile communications is towards converged next generation broadband networks, which offer abundant bandwidth and seamless integration of both fixed and mobile networks. On the other hand, there is an increasing interest in services such as video-on-demand, television and other content services while on the move. Some questions that are still unanswered are what will be the likely scenario of competition in infrastructures as a result of a migration to next generation networks (NGN), would there be differentiated prices for different services, would the demand for interoperability increase, how can we offer content services worldwide?

The technological landscape has changed significantly during the past decade. New communication technologies, new media, the Internet and devices carrying new functionalities are expected to meet consumers' demand for seamless, simple and user-friendly communication means providing access to an extended range of services and content.

Next generation or 4G is defined as a completely new fully IP-based integrated system of systems and network of networks achieved after convergence of wired and wireless networks as well as computers, consumer electronics, and communication technology and several other convergences that will be capable to provide 100 Mbps and 1 Gbps, respectively in outdoor and indoor environments, with end-to-end QoS and high security and low power consumption, offering any kind of services at any time as per user requirements, anywhere with seamless interoperability, always on, affordable cost, one billing and fully personalized [2]. This is shown in Figure 21.1.

Traditionally, audio, video, data or voice communication services were accessed through different ICT networking infrastructures and distinct terminal devices. These included PCs connected to the Internet, TVs picking up broadcast signals, telephones connected to copper/fiber local loops, or mobile devices connected to wireless networks. However, 4G must follow NATO NIAG's demands on coexistence between military and civil wireless services. This is particularly important in view of the fact that the EU is opening up its defense market that until now had been largely closed, and has adopted a Code of Conduct requiring defense contracts to be awarded on the basis of Europe-wide invitations to bid [3]. The new rules, which at first were non-binding, apply as of July 2006, and compliance is being monitored by the European Defense Agency (EDA), whose mission is to develop joint

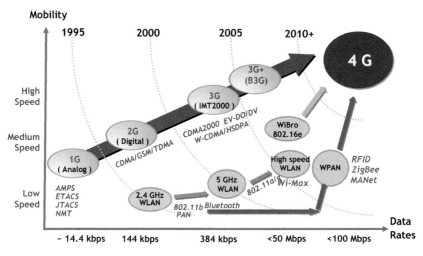

Figure 21.1 The road towards 4G [2]

defense capabilities, cooperate on armaments, and liberalize the market as well as to promote defense-related research and development. Particularly, in the area of research and technology, the overall approach to looking at the research landscape taken by the European Commission and the EDA is becoming ever more clear and European.

Further, since World War II, military superiority has been based on a techno-logical advantage, and technology is now even more important in the unpredictable security challenges the world faces. However, many of the new technologies most critical to defense are emerging in the commercial consumer sector, and defense access to these technologies is limited.

Therefore, coexistence of future technologies in the frames of NGNs must include also the requirements for coexistence of technologies across different societal sectors while providing an appropriate to each of these sectors degree of security. This newly elevated context of coexistence of technologies will depend on solving a number of challenges related to the following areas:

- Spectrum management;
- Standards management;
- Emergency and rescue services handling;
- Communications and navigation architecture;
- Technology and operations integration;
- Security, privacy and trust.

This Chapter discusses the importance of convergence of navigation and ICT communication for the adoption of NGN and the ultimate migration towards 4G. The other message of this Chapter is to outline a roadmap for ensuring coexistence of technologies that are equally important for the commercial and military sectors.

One very important question that arises here is how we can achieve technical feasibility between the technologies supporting applications in these two different sectors, and what would be the business/security risk to support military applications with commercial communication system technologies.

The Chapter is organized as follows: Section 1.2 discusses the challenges of convergence, in general. This section identifies convergence as the main factor for change for the ICT sector and the society at large. Section 1.3 discusses the specific characteristics of navigation and communication, the services that can be offered by both systems and by an integrated approach, and the emerging technologies making such a convergence possible. Finally, Section 1.4 identifies the challenges of the two areas of navigation and communication, also in relation to security, and how these can be overcome. Section 1.5 concludes the Chapter.

21.2 Convergence of Technology

The various networking environments and technologies have also historically been developed on the basis of very different business models, with different players at each level of the value chain. In the context of these models a service is ultimately coupled with a network infrastructure (e.g., a mobile phone call is conveyed primarily over a dedicated mobile infrastructure, a radio station broadcasts audio content, an e-mail is sent over the Internet).

Today's technology convergence drive is radically changing this picture, with the result that progressively, a complete separation is taking place between the underlying networked infrastructure and the services/applications it can deliver to the user, be it at home, at work or from a mobile device. The broad range of industries involved in convergence at the various levels includes IT (hardware and software), consumer electronics, electronic communications, broadcasting and content providers including media, and large Internet companies. These industries have different backgrounds but they are finding themselves competing in new markets as a result of common platforms, networks and services with similar functionalities. Often completely different infrastructures like GPS and 4G are utilized at the same time, thus bringing conflicts between separate regulation regimes.

Convergence is bringing about industrial changes both at the horizontal level, whereby traditionally separated industries compete with each other, and at the vertical level, whereby new partnerships emerge bringing about the need for new business models and sometimes trends towards vertical integration. Convergence is blurring the boundaries between markets and strengthens vertical links. Communication services, delivery devices and media content are increasingly interrelated: economic distortions in one sector may easily spill over to another sector. The traditional telecom regulation is challenged in this new field.

Globalization is interrelated to convergence and is the existence of many communication, computer and information systems, each having different functions, capabilities and challenges (see Figure 21.2).

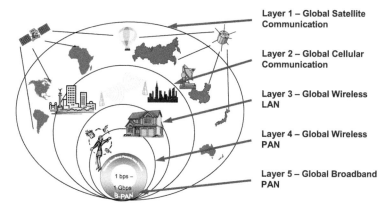

Layer 1 – Global Satellite Communication

Layer 2 – Global Cellular Communication

Layer 3 – Global Wireless LAN

Layer 4 – Global Wireless PAN

Layer 5 – Global Broadband PAN

Figure 21.2 Global communications through technological convergence

Most of these systems communicate constantly with each other (e.g., monitors, controllers and routers) or for the duration of specific transactions (e.g., phone calls, document transfer, video and audio streaming, and so forth). This means that governments, businesses, households, and individuals are dependent on the global communication system to reliably perform all required interactions and to be robust against security attacks or potentially damaging changes in the communication or operational environment, in order to support different activities. If the security postures of the global communication system are imbalanced in one part, this could pose problems for the overall communication process and anyone involved in it. Users depend in their every day social and business interactions on different types of relationships, such as authorities, family, friends, business associates, or third parties, to alert them about possible/pending threats to their property or person.

One problem of the global (ubiquitous) communication system as described above is how to establish a secure, dependable and trustworthy relationship-based infrastructure that provides the required mechanisms and protocols for ensuring cooperative security of the various communication processes and users.

21.2.1 Next Generation Communications

Research activities in the area of next generation communications are seen as the basis to deliver ubiquitous and converged network and service infrastructures for communication, computing and media [5]. This entails overcoming scalability, flexibility, dependability and security bottlenecks, as today networks are primarily static and able to support only a limited number of devices and applications. Such new infrastructures will permit the emergence of a large variety of business models, involving a multiplicity of devices, networks, providers and service domains. The current research efforts towards the next generation of converged communications are building on the results of research work towards the adoption of beyond 3G systems and systems that can ensure broadband communications for all users regardless of geographic location.

Mobile and wireless systems beyond 3G realize the vision of "optimally connected anywhere, anytime" supported by all system levels from access methods and networks to service platforms and services. Research work has characterized Systems beyond 3G as a horizontal communication model, where different terrestrial access levels and technologies are combined to complement each other in an optimum way for different service requirements and radio environments [4]. Further work towards next generation communications requires long-term developments for advanced physical layer techniques in broadband communications, fixed and mobile terminals, and 4G evolutions with a potential convergence between different technologies. In particular, a migration to a next generation of communication systems requires:

– A consolidated approach to technology, systems and services, including location-based services, notably in the field of future standards (e.g. for access) and in international forums (WRC, ITU, 3GPP-IETF, ETSI, DVB . . .) where the issue of systems beyond 3G is addressed;

– A consolidated approach regarding the spectrum requirements (terrestrial and satellites) in the evolution beyond 3G and a clear understanding of the novel ways of optimizing spectrum usage when moving beyond 3G.

A generalized and affordable availability of broadband access to users, including those in less developed regions, peripheral and rural areas, requires optimised access technologies, as a function of the operating environment, at affordable price and allowing for a generalized introduction of broadband services on a global scale and in less developed regions.

The long-term evolution in communications is expected to lead to the consolidation of access platforms, allowing users to seamlessly access content and services through a variety of fixed and wireless networks.

Convergence is pushing towards an environment that requires new investment in infrastructure able to support the delivery of rich services, applications and content. Investment needs to take place both in the core networks as well as the access level. From this point of view, the main regulatory challenge concerns the adequacy of rules in providing sufficient incentives for long-term investment to take place. Factors to be taken into account are the migration to IP networks, the need for more capacity for the final user, the importance of sustainable competition at the access level, the demand for availability of content on fixed and mobile platforms and of the development of new business models for the delivery of advanced services. Policy should aim at the creation of a favorable environment for investment to take place. This means legal certainty and sustainable competition.

21.2.2 Growing Demand for Radio Frequency Spectrum

Expanding satellite and multimedia applications are driving growing demand for radio frequency spectrum [6]. The number of satellites operating in the 7–8-GHz band commonly used by geosynchronous orbit (GEO) satellites has been increasing. Satellite operators now spend about five percent of their time addressing frequency interference issues.

The growth in military bandwidth consumption has also been dramatic. For example, the US military used some 700 Mbytes per second of bandwidth during Operation Enduring Freedom in 2003, compared to just 99 Mbytes per second during Operation Desert Storm in 1991[7]. In Europe, within the frequency band 20–108 MHz the common military tuning range is 30–87.5 MHz, however, some equipment types use the lower (20 MHz) and upper (108 MHz) limits, regulated on a national basis. The harmonized military bands are:- 30.30–30.50 MHz; 32.15–32.45 MHz; 41.00–47.00 MHz; 73.30–74.10 MHz; 79.0–79.70 MHz. When providing for additional requirements, further blocks of frequencies should be spread out over the whole common military tuning range in order to supply frequencies for frequency hopping equipment and to support a larger force (corps size, three divisions) [8].

Spectrum management is therefore crucial in order to provide for sufficient bandwidth to all users within the global communication society. If we consider the need for technological convergence across different societal sectors, spectrum management must provide solutions for interference-free access to use of spectrum for communications that support launch, orbiting, navigation, telemetry, control and sensor activities, as well as use of spectrum for interference-free access to personalized services, including entertainment services.

Another demand on frequency allocation comes from the dual-use of navigation and communication technologies (NAVCOM). A NAVCOM scenario will imply that the same modules of the NAVCOM system will be used for both navigation and communication, and therefore, frequencies will also experience dual use. This can only be possible if both technological (e.g., coexistence) and regulatory issues are solved with success.

With the emergence of cognitive radio, the problem of spectrum sharing gets new possibilities but also challenges, especially in relation to the adoption of dual use technologies. As an example: Shared use of spectrum where orthogonal modulation schemes are used. One of the major challenges lying ahead is to ensure security of the dual use communication process and system.

21.2.3 Security Requirements

Citizens and consumers can derive a wide range of benefits from convergence, going from easier access to a great variety of information, to a greater capacity to create and develop their own content. However, many people are still reluctant to use information society and media services, due to concerns regarding the safety of their children, the integrity and stability of their computers or the abuse of their personal data.

Technological and business developments resulting from convergence, such as, Internet access on mobile phone, RFID, digital rights management (DRM) or location-based services raise a number of new concerns regarding protection of minors, consumers or human dignity, security and privacy.

All these issues need to be addressed in order to provide consumers with the level of protection they are entitled to expect from any environment. The security requirements become even more important when convergence extends to the convergence of technologies that would support services in both the commercial and military sectors. Such an evolution demands full-proof mechanisms for consumer protection on one side, and military security on the other. There are challenges to face regarding unwanted access to unsuitable, extreme, adult or harmful content and to risky communications. Securing network infrastructures and services is also a major concern for economic players.

With the development of new services and increased numbers of business transactions carried out over next generation networks (NGNs), information society and media service providers' losses resulting from security breaches could become considerably higher. The infrastructure of NGN/IP networks is packet-based and multilayered, with open, distributed architecture and no embedded security mechanisms. However, these networks are used for the transmission of high-profit services such as voice, e-commerce or financial transactions.

In a technological environment that is constantly evolving, with the development of new services and content delivery platforms, the main challenge is to be able to identify any new security- and network resilience issues, in order to give it the appropriate technical or regulatory remedy.

21.3 Navigation and Communication Converged Architecture for a New Dimension of Services

Current technological trends in terrestrial telecommunication sectors show that seamless integration and high data rate are the two key elements to achieve the ITU-R vision of "optimally connected anywhere anytime". Therefore it appears that exploring different options for higher data rate radio interfaces and integrating them under one unified core network are two major tasks of the people working in the telecommunication arena.

In a converged NAVCOM architecture, navigation technologies (like GPS) enable accurate timing, positioning, and navigation capabilities on a worldwide basis, while communication technologies (like PCS satellites) enable worldwide communications. Related to dual use, a converged NAVCOM architecture will be used both for military and civilian service provision.

However, there are several issues, which need to be resolved for a successful adoption of the converged NAVCOM architecture [9]. These can be summarized as follows:

(1) Can a cost-effective NAVCOM system be developed that meets all of the requirements of both military and civilian users?
(2) Why should the military consider dual-use?
(3) Why should the civilian sector consider dual-use?

(4) Can dual-use be developed without compromising military performance, availability, and security?

(5) How does dual-use benefit the military as well as the civilian sector?

(6) Why should and will navigation service and communication services be integrated on the same satellite system?

Dual use of navigation and communication technologies will put even more stringent requirements on the quality of service (QoS) in addition to security. In fact, both will be equally important and there should not be a trade off of security to provide QoS or vice versa. Therefore both can go under the same umbrella and be referred to as quality of service and security (QoSS).

21.3.1 Scenarios for Services over Converged NAVCOM Systems

21.3.1.1 Satellite Services

The provision of satellite services is based on the data available from the currently operational satellite systems and the terrestrial telecommunication systems. Current trends in the satellite industry show that the required technology for the future satellite systems will be supporting both mobile and broadband. Some implied research issues relate to satellite constellation, payload technology, radio-interface, inter-working, networking technology, user equipment technology).

Figure 21.3 shows an architecture for providing services to users over the converged satellite, mobile and broadband infrastructure.

There are a number of service profiles that can be associated with the scenario of Figure 21.3. Namely, services can be provided to the home, office, or to remote users. A number of applications that can be mentioned here are e-learning, telemedicine, e-payment, games and other services that may improve the comfort of users located far from large urban areas. The scenario shown in Figure provides high level of connectivity based on a hybrid network combining a satellite segment and terrestrial wired (Ethernet) and wireless (WiFi, WiMax) access. The different types of interconnections can be used simultaneously and are transparent to the end users [11]. Also there [11], a cost analysis was made for the scenario and it was reported that the price would be two times lower while the capacity was three times bigger. The main cost decrease came from the installed satellite terminal. A shared satellite access would bring the price further down as it would be shared among all users. It can be concluded that satellite and other communication technology are complementary, and therefore, expansion of access to communication services via satellite is a viable solution. However, it must be mentioned that satellite solutions are expensive and limited in their performance and evolution potential, while their costs rise steeply with the demand for richer services. A 2-Mbps-both-way satellite service at a five year cost of €140,000 would appeal only to very specialized customers [11]. Key issues in the consideration of satellite broadband will be the cost trends for satellite

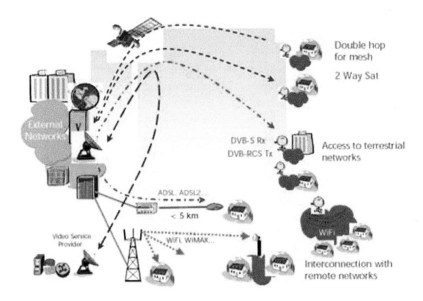

Figure 21.3 Converged satellite, mobile and broadband architecture for the provision of services [10]

services themselves and how newer technologies challenge satellite for providing cost-effective services for a multitude of users.

21.3.1.2 Data Telemetry

Data telemetry is another application that can simplify and speed the acquisition of critical information from remote locations. Remote telemetry systems, for example, take advantage of the latest wireless communication systems to create integrated environmental data collection solutions. The most effective wireless communication technologies for environmental data include cellular, radio and satellite telemetry. Bouncing an RF signal off a satellite is one of the best options for remote installations, especially in locations where no other reliable RF or telephone coverage is available. Satellite is also recommended when infrastructure cost (such as the need to use repeaters for a non-satellite system) is an issue. There are two main types of satellite systems that offer remote environmental monitoring type applications: low-Earth orbit (LEO) and GEO. Radio telemetry is one method of transmitting the data. Licensed and unlicensed radio telemetry may require line-of-site (LOS) between the monitoring system and the base station. Antenna positioning, gain, tuning, ambient frequency, noise, atmospheric conditions and terrain are important variables affecting range, however, transmissions can be repeated if interference is present. A summary of the typical point-to-point transmission range of different telemetry modem types is shown in Figure 21.4.

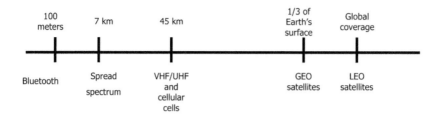

Figure 21.4 Typical maximum point-to-point transmission range of respective telemetry modem type [12]

21.3.1.3 Public Safety

Public safety applications are normally used by police, fire, search and rescue in response to emergency and medical situations. The primary service of a public safety system is mass-alert from an entitled authority to the population. This would require that some decision centers are set up and provided with reliable communication link. Normally such centers are located in big cities with a fast cable connection. However, in the case of major disasters, the terrestrial infrastructure may be down and alternative communication means must be available.

Two-way preemptive communications can also be envisaged between rescue teams and a rescue center, for example. Such communications would engage richer content and would require high QoS.

The domain of public safety has recently extended to the vehicles. The integration of information and communication technologies (ICT) in vehicular environments enables new kinds of applications and creates new technological challenges. Dynamic network topology, unreliable network links, and moving terminals are some of the challenges of a successful public safety network.

Next generation communication systems must have capabilities for the provision of stand-alone location information that can be used as input for user and system-side applications, e.g., emergency services [13]. Further, such systems shall aim at achieving user emergency call requirements. Concerning accuracy requirements a next generation system shall at least meet the requirements for the unified E-112 system agreed by the EU [13]. However, currently no agreed European requirements exist, just recommendations by [14]. The requirement that a next generation system should maintain compatibility with existing location information mechanisms (GPS, GALILEO) [13] concerns an "add-on" for a better location determination performance for, e.g., location-based services or more advanced handover.

A location determination system function can be used to handle location requests, originated either from the user terminal (UT) or from some service in the network [15]. When a location request is received, the location service support makes sure that user's privacy is not violated and/or checks that using the location service is allowed and charged for the specific user, and then it initiates the positioning process. Generally, the location determination (LD) can be done within the UT using performed measurements and information sent by the location service support function, or within the location service support using measurement reports sent by the UT and/or BSs and relay nodes involved in the location estimation process. If appropriate capabilities of the UT are available, also the global navigation satellite system (GNSS) based support from GPS and/or GALILEO can be included.

In [16] general positioning methods were introduced and described in a more abstract way. Descriptions were given for

- Time of arrival (TOA);
- Time difference of arrival (TDOA);
- Received signal strength (RSS);
- Angle of arrival (AOA);
- Cell ID.

However, different QoS requirements for the different applications require the flexible and scalable radio access network as it is proposed by [6], [13]. Applications that can use the location of the UT could be: point-to-point navigation, emergency call handling, location-based handover, or location-based service provisioning. In order to cater for different QoS demands, the location information should also be scalable in order to provide appropriate accuracy for the involved UTs with different profiles [17].

Generally, the location estimation can be done within the UT using measurements and information sent by the location service support function, or within the location service support using measurements sent by the UT and/or base stations (BSs) and relay nodes involved in the location estimation process. The best stand-alone based performance can be obtained by including timing measurements of the UT. In-band timing measurements in a cellular network based positioning are usually based on TDOA measurements [17]. They are based on the idea to find the starting point (TOA) of the incident orthogonal frequency division multiplexing (OFDM) signals to estimate the distances between the UT and the BSs using the included pilot sequences [18]. If the GNSS based positioning information is included (GPS, GALILEO), the performance can be further improved. Tracking algorithms for the solution of the navigation equation in the dynamic case can be applied to obtain accurate TDOA timing information.

Usually, the user with its UT is moving around a certain track in different scenarios. Clearly, there are certain correlations between the positions of the UT over time. This information can be integrated in the overall position estimation process and can help to improve the estimates in average.

The Kalman filter (KF) is a flexible tool for providing such positioning estimates in the context of tracking applications. However, the standard KF just performs optimal if several criteria on, for example, linearity or Gaussianity, are fulfilled, which is usually not the case in practical applications. Nevertheless, even

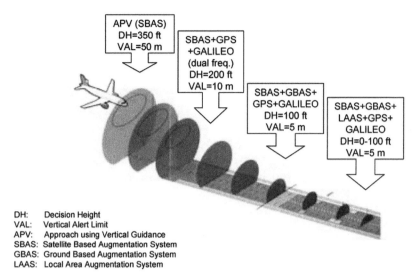

APV (SBAS)
DH=350 ft
VAL=50 m

SBAS+GPS
+GALILEO
(dual freq.)
DH=200 ft
VAL=10 m

SBAS+GBAS+
GPS+GALILEO
DH=100 ft
VAL=5 m

SBAS+GBAS+
LAAS+GPS+
GALILEO
DH=0-100 ft
VAL=5 m

DH: Decision Height
VAL: Vertical Alert Limit
APV: Approach using Vertical Guidance
SBAS: Satellite Based Augmentation System
GBAS: Ground Based Augmentation System
LAAS: Local Area Augmentation System

Figure 21.5 Future aircraft landing system [2]

if these conditions are not fulfilled completely, the KF gives reliable and robust estimates [15].

21.3.1.4 Airport Services

Air traffic control (ATC) is a service provided by ground-based controllers who direct aircraft on the ground and in the air (see Figure 21.5). A controller's primary task is to separate certain aircraft — to prevent them from coming too close to each other by use of lateral, vertical and longitudinal separation. Secondary tasks include ensuring safe, orderly and expeditious flow of traffic and providing information to pilots, such as weather, navigation information, etc.

Air traffic control can be optimized by ensuring access to real-time information related to aircraft position and four-dimensional trajectory intention, precise real-time and forecast weather information, and airport surface and terminal conditions. Technologies ensuring this but still immature are in essence converged NAVCOM technologies: the four-dimensional trajectory air / ground data exchange, aircraft-to-aircraft position information, sophisticated flight data and ATM systems, integrated airport and airport surface information and management systems, sophisticated weather forecasting capabilities and robust data link capabilities.

21.3.1.5 NAVCOM Services for Sustainable Growth

NAVCOM convergence can provide a new push in ICT research involving major stakeholders such as the automotive and transportation industries, equipment suppliers, the telecommunications industry, motorway, road infrastructure and fleet operators, utility providers, public authorities, civil protection and service providers.

NAVCOM convergence will promote the development of intelligent vehicle systems that offer a higher degree of accident prevention through improved driver-warning strategies, hazard detection, actuation and sensing including sensor fusion and sensor networks, as well as the integration of independent safety systems and their interaction with the driver, as well as solutions for efficient and secure freight transport.

For the provision of the above services, it is necessary to integrate a number of advanced technologies, such as low-cost GNSS receivers, software defined radio technologies (SDR), high-accuracy hybrid positioning systems combined with dynamic navigation services, semantic web and multi-agent technologies, as well as technologies such as RFID and smart tags in combination with advanced sensors, communication and mobility management systems.

NAVCOM convergence will help integrate environmental monitoring and management with an enhanced capacity to assess population exposure and health risks, to report to and alert targeted groups and to organize efficient response.

Further, NAVCOM convergence supports communications in a decentralized fashion. NAVCOM services can be a solution in situations where a number of peers need to exchange information without existing communications infrastructure. For example, cars aiming at coordinating a possible encounter at a road crossing should rather establish peer-to-peer links than relying on cellular networks. The latter would have problems handling the multitude of information exchange in city areas with dense traffic, or they may be unavailable in tunnels or underground car parks. Ad hoc networks, on the other hand, can be set up locally by a number of cars wherever needed. The cars are spacious enough to be equipped with multiple antennas for the sake of high network capacity. The use of multi-element antennas along with adequate signal processing greatly enhances the spatial channel reuse in decentralized wireless networks.

Another benefit form the NAVCOM convergence are the personalized applications increasing the comfort of the driver and the passengers. Enhanced information services, navigation system or personalized vehicle settings. Examples are Internet access, personalized information on the road, and the like. Convenience applications extend the reach of mobile services to the vehicles. In addition, new services can be created based on the context information available in vehicular environments, such as group navigation (follow me), gaming (for the passengers), or location-based promotions.

21.3.2 Enabling Technologies for Converged NAVCOM Systems and Dual-Use Applications

21.3.2.1 Technological Advances in the Design of Sensors and Processors

Technological advances in the design of sensors and processors have facilitated the development of efficient embedded vision-based techniques. Distributed algorithms can provide more confident deductions about the events of interest or reduce ambiguities in a view caused by occlusion or other factors. Because they operate in real time, a variety of smart environment applications can be enabled based on the development of efficient architectures and algorithms for distributed vision networks. Distributed smart cameras combine techniques from computer vision, distributed processing, and embedded computing. As a non-invasive technology, imaging plays an important role in mobile sensing devices. Multiple communicating cameras viewing the same scene from different viewpoints can be combined into a high performance surveillance system. Wireless smart cameras challenge the hardware for low-power consumption and high image-processing performance. These are also drivers for typical dual-use applications such as adaptive optics, laser communications, target tracking and image processing. To make these possible, research must focus on new real-time algorithms for adaptive filtering, prediction, and system identification with improved efficiency and numerical stability for the large numbers of channels and high filter orders.

21.3.2.2 Femtocells

A femtocell is a very small footprint mobile base station using wide area network radio access technology, such as UMTS, that would be deployed in a single-family home or multi-dwelling unit building to serve the voice, data and potentially video needs of four to eight users. In a similar way to the microcell or picocell, it is designed to serve the mobile needs of a larger number of users. Femtocells could give mobile carriers entry to the home with services that would allow consumers to use existing 3G handsets, rather than having to buy a new fixed-mobile converged handset.

Femtocells can help carriers solve problems of both coverage and capacity which are applicable to all standards. Femtocell access points offer cellular carriers the opportunity to address fixed-mobile convergence markets with a highly attractive and efficient solutions. They provide savings on backhaul costs, improve in-building coverage, reduce churn, promote migration and provide a platform for operators to build an effective delivery system for triple and quadruple play services [19]. Still, some of the challenges that need to be solved are the provision of access to several cellular networks from a single femtocell, the design of algorithms for automated network planning with femtocells, the provision of access to special services (e.g., TETRA), the integration of different existing technologies and enabling adding new ones in the future, and the robustness of the node. Especially, the last challenge is very important for the adoption of converged services for dual use.

21.3.2.3 SDR and Reconfigurability

SDR, reconfigurability, and portability are another enabling technology for converged NAVCOM systems. Reconfigurability allows that the same platform can host multiple waveforms with the same platform services, while portability suggests that the same waveform can be implemented on multiple platforms with the same waveform software (see Figure 21.6).

These two concepts enable the user with a portable-all-in-one radio device. An SDR-based platform provides flexibility through a heterogeneous multiprocessor

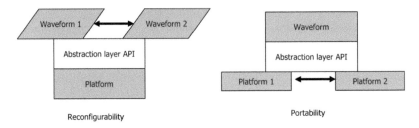

Figure 21.6 Reconfigurability and portability concepts

that incorporates different types of software programmable processors and reconfigurable hardware components, such as field programmable gate arrays (FPGA) [20]. Advanced SDR receivers will strengthen the use and role of satellite navigation in the civil and military society functions and services because they allow for flexibility, adaptability and dynamicity for existing and new applications and services.

21.3.2.4 Cognitive Radio

The idea behind cognitive radio is to access the channel in intelligent ways. For instance, channel sensing can render information which helps to find free bandwidth resources, so-called spectrum holes. Cognitive channel access policies could facilitate dependability and high throughput in peer-to-peer networks. Conventional channel prediction methods may not be suitable in this context. Novel methods need to be devised for predicting the channel usage in uncoordinated peer-to-peer networks. Candidates may be estimation algorithms for non-linear systems, machine learning or other learning algorithms.

Cognitive radio is expected to provide a means to utilize frequency spectrum in more flexible and dynamic manner. One of the standardization related to such a dynamic spectrum use is IEEE 802.22 [21]. IEEE 802.22 devices can utilize the frequency spectrum currently allocated to TV broadcasting while avoiding harmful interference to TV receivers. The required techniques to achieve interference awareness, such as cooperative spectrum sensing, decision to start transmission, and all the system parameters are defined in the standard. This standard basically aims at increasing spectrum efficiency, which can result in the increase of available bandwidth for wireless communication systems. Most of the works related to dynamic spectrum usage based on cognitive radio have been trying to achieve similar goal, i.e., the increase of the available bandwidth/data rate for wireless devices.

21.3.2.5 Cooperative Communications

The idea of improving the performance of communication networks through cooperative behavior has recently attracted much interest. Methods have been proposed for single-antenna devices which increase the diversity order and thereby reduce outage probabilities. Similar cooperative communication techniques are also conceivable for decentralized networks of devices with multiple antennas. These techniques could help to keep outage rates limited and thereby contribute to system dependability. Cooperative communications are also very important in the scope of next generation communication systems.

21.3.2.6 Remote Sensing Systems

The expected development of technique and technology will lead within the next decades to establishing an autonomous, global NAVCOM system with integrated communication, positioning and navigation capability which has one or more central illuminators together with a synchronized fleet of both airborne and spaceborne receivers which enable global coverage and continuous availability. A first major step could be the use of GPS or GALILEO satellites as transmitters which in a later step could be especially equipped (exemplary during maintenance and exchange) with dedicated wide band and high power SAR transmitters where the comparably small GPS bandwidth of 10 MHz is embedded [22].

21.3.2.7 Space Technology

Space technology also reaches remote and uninviting territories, the seas and the airspace above. Mobile broadband communication via space opens the way to safety measures such as live CCTV monitoring of passenger aircraft from the ground, and resilient nationwide traffic automation. On the strategic level, the global reach of unencrypted direct-to-home satellite broadcasting without interfering state censorship plays a central role in promoting international understanding and security. Space engineering faces unique weight and size restrictions, energy constraints, extreme environmental conditions and highly challenging sensor and data transmission requirements. These have consistently worked as key drivers for broader technology innovation and continue to play a central role in the further advancement of information and communications technology as well as new materials.

21.4 Challenges of Converged Navigation and Communication Systems

A converged NAVCOM system will comprise various communications and remote sensing systems. It is necessary that there is availability of equally protected frequency bands to ensure their successful implementation. Therefore, spectrum management and adequate spectrum policies are a key prerequisite for the adoption of services delivered over converged NAVCOM systems.

The present communication and navigation systems have small and cheap user units (to a wide extent standardized) and more or less centralized transmitters. Hence, it is necessary to learn from these already existing systems and to take over respective technologies, techniques and even components. This predestines in some cases the frequency range, L-Band for example, by using GPS or GALILEO components as well [22]. The dual use of the same frequencies for radar and other services will be indispensable as well as design to cost and use of existing technologies, products and competence. This implies the need for use and application of electronic components gained with other microwave communication and navigation programmes.

Convergence of communications allows for parallel use of dedicated navigation and communication systems, which evolves NAVCOM systems into situation-aware multiple-use, single infrastructure systems. It allows that in case of a degradation of one segment, take-over of functionality by other segments can be programmed or self-learned and that safety-relevant redundancy can be allocated on-demand. However, some of the challenges to be solved are related to reconfigurability of the backhaul network and radio access, cross-network detection of unexpected situations in network operations, and cross-network optimization of radio resource use.

A large number of challenges to be solved before a successful adoption of converged NAVCOM services relate to security. As most of the NAVCOM services will be requested by critical infrastructure sectors (e.g., health, air-traffic/air-passengers handling, vehicular infrastructure and pedestrian safety, etc), ensuring control over all communication networks is needed because it enables prioritizing services, blocking certain traffic and/or access to networks from given locations, if it turns out to be necessary.

Converged NAVCOM services will rely on technologies ensuring that the underlying data is complete, accurate and not corrupted. This demands an improved information system to ensure that bad data is not widely disseminated. It is vitally important, to provide security solutions ensuring that the validity and correctness of underlying data is verified.

For a successful adoption of NAVCOM system location-determination techniques must be further refined. Several factors have an impact on the quality of the location determination estimates. Some of them can be influenced by the UT or network; some of them just depend on the environment and the network topology [23]. For location determination purposes, it is necessary that measurements to/from at least three BSs are performed. To get a strong signal from the serving BS is usually not a problem. But to do accurate measurements with signals received from neighboring cells can be difficult because the received signal strength (RSS) is very weak. Additionally, a next generation communication system can work with frequency reuse of one [24] resulting in interference problems degrading the performance. The following factors will have impact on the positioning performance:

- Quality of the measurements: This depends on how often, how quickly, and how accurately the measurements can be performed in the system. The general system parameters (e.g., bandwidth, signal shaping, pilot structure) influence the quality. For example, if the bandwidth for different deployment scenarios is between 50 Mhz (wide area) and 100 MHz (local area) it would yield an equivalent accuracy between 6m (WA) and 3m (LA) if synchronization within one chip can be achieved [24]. However, besides weak signals from out-of-cell BSs and inter- and intracell interference, disturbing effects like multipath and especially non-line of sight (NLoS) propagation can limit the positioning performance. Figure shows simulations for the

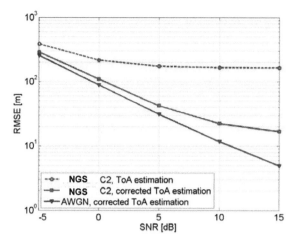

Figure 21.7 Link-level ToA positioning accuracy for a next generation system (NGS) [23]

expected root mean square error (RMSE) for ToA estimation by adapting Minn's algorithm [25] for link-level synchronization.

– Geometrical constellation: The position of the UT related to the position of the involved BSs has also an impact of the positioning performance, i.e., the geometry of the BS network has a direct influence on the performance.

– UT capabilities: This includes different configurations of the UTs (e.g., GNSS capabilities) and their multimode capabilities.

Location determination can be a valuable add-on in assisting mobility management of users in next generation systems.

12.5 Conclusions

Communication and navigation technology has now moved from the sphere of early research, technology development and isolated specialist applications to widespread, routine operational use: communications, navigation (positioning and timing) and earth observation, including weather monitoring. In Earth observation, only satellites can capture large-scale phenomena such as weather systems, ocean currents, variations in sea levels or soil humidity. They are also unique as legitimate tools for global information gathering without the need for prior consent and knowledge by other countries.

Satellites are also useful means for ensuring communications to remote geographic locations (e.g., fleet vessels, locations with poor cable infrastructure).

ICT markets are shaped by standards. Convergence of navigation and communication will lead to new standardization activities that will launch new markets, services, and products. The future of converged NAVCOM is dictated by the application requirements of the users and the expected progress of technique, technology and economical factors.

Dual use for both, military and civil applications would require a better relationship between investments and outcome and the promotion of new applications. The dual-use nature of NAVCOM technology and the strategic advantages its control offers demand an inclusive approach to space and communication policy both that views civil and security space as two sides of the same coin, at least for the purpose of identifying objectives and priorities. This dual nature also poses a constant challenge to the direction of NAVCOM technology activities. It has to be remembered that GPS is an example of military and civil use.

Finally user friendly and ingenious MMIs are necessary for NAVCOM success, hence the new converged services will end just as two separated devices sharing the same battery and display!

References

[1] i2010High Level Group, "The Challenges of Convergence," A Discussion Paper, European Commission, December 2006.
[2] Prasad, R., and Lauridsen, O. M., "Convergence of Navigation and Communication (NavCom) Towards 4G," An AFCEA International Symposium on AEROSPACE TECHNOLOGIES and APPLICATIONS for DUAL USE, Rome, Italy, September 2007.
[3] German Defense Industry Committee, Position Paper, August 2006.
[4] FP6 EU Framework Programme, www.cordis.europa.eu/ist.
[5] FP7 EU Framework Programme, http://cordis.europa.eu/fp7/ict/future-networks/home_en.html
[6] International Telecommunications Union (ITU) available at http://www.int.org.
[7] Clinton, B., "Clinton to Rename DARPA, Expand Emphasis on Dual-Use Technologies," 2005, at www.accessmylibrary.com/coms2/summary_0286-5492447_ITM - 24k.
[8] EU-Footnotes, Doc FM(01)123 Rev 1, August 17, 2001.
[9] Esposti, R., Di and Jonahsen, J., " Dual-Use Personal NavCom System," *IEEE*, 2000.
[10] IST Project 026590 SATSIX, Deliverable D100_4, "Satellite Network Requirements," January 2007, available at: http://www.ist-satsix.org
[11] IST Project 026590 SATSIX, Deliverable D100_3, "Collective Access Scenario," Public Report, July 2006.
[12] http://www.stevenswater.com/telemetry_com/index.aspx.
[13] IST Project WINNER II, Deliverable 6.11.4, "WINNER System Requirements," November 2005 at www.ist-winner.org.
[14] Coordination Group on Access to Location Information for Emergency Services (CGALIES), "Final Report: Report on Implementation Issues Related to Access to Location Information by Emergency Services (E112) in the European Union," http://www.telematica.de/cgalies/, February 2002.
[15] IST Project WINNER II, Deliverable 4.8.3, "Integration of Cooperation on WINNER II System Concept," November 2007, www.ist-winner.org.
[16] IST Project WINNER II, Deliverable 4.8.1, "WINNER II Intramode and Intermode Cooperation Schemes Definition," June 2006, www.ist-winner.org.
[17] IST Project WINNER II, Deliverable 4.8.2, "Cooperation Schemes Validation," www.ist-winner.org.
[18] Mensing, C., et al., "Location Determination Using In-Band Signaling for Mobility Management in Future Networks", *Proceedings of the IEEE International Symposium on PIMRC*, Athens, Greece, September 2007.
[19] ABI Research, Report on "Femtocell Market Challenges and Opportunities," 2Q, 2007.

[20] Hoebner, M., et al., "Exploiting Dynamic and Partial Reconfiguration for FPGA's: Toolflow, Architecture, and System Integration," *Proceedings of 19th ACM Annual Symposium on Integrated Circuits and System Design*, August 2006, pp.1–4.

[21] IEEE 802.22 available at http://www.ieee802.org/

[22] Keydel, W., "Considerations Towards the Future of Space-Borne SAR Systems," *Proc. of URSI'05*, 2005.

[23] Mihovska, A., et al., "Policy-Based Mobility Management for Heterogeneous Networks," *Proceedings of IST Mobile Summit'07*, Budapest, Hungary, July 2007.

[24] IST Project WINNER II D6.13.7: "Test Scenarios and Calibration Cases Issues 2," Deliverable, IST-4-027756, December 2006, http://www.ist-winner.org.

[25] Minn, H., et al., "A Robust Timing and Frequency Synchronization for OFDM Systems," *IEEE Transactions on Wireless Communications*, Vol. 2, No. 4, July 2003, pp. 822–839.

Chapter 22

Innovative Satellite Applications for Homeland Security

L. Pasquali

Telespazio

22.1 Introduction

Homeland Security is the ultimate process, representing the most complex system of systems for human imagination. All the sensitive activities, critical infrastructures, transportations, noticeable sites, green blue and air borders, mass events, each single human life have to be protected against natural and anthropic threats. The risk of terrorist attacks has to be managed and minimised in a sustainable way, with minimum changes in people way of life, optimising the investments and limiting the presence of invasive infrastructures.

In addition terrorists modify with time their targets and the Homeland Security Process shall prevent if possible, decide the soonest and fast react to any kind of threat, even unsuspected.

Satellite systems are therefore substantial contributors for the intelligence data gathering and elaboration, communications and reaction phases which are the basic elements of the Homeland Security adaptive process.

Satellite systems can supply the maximum performance, delivering integrated earth observation and localisation/communications services over wide geographical areas, with the lightest possible infrastructure: they are therefore the necessary framework of any feasible Homeland Security Process, to be integrated in an interoperable scenario with the other local surveillance and communications systems.

The awareness of the importance of the above considerations is well demonstrated by the GMES programme, which is defining an integration strategy between satellite and terrestrial (including air and maritime platforms) systems.

In this presentation a few innovative applications are described with some detail:

- maritime applications, critical infrastructure monitoring, illegal immigration surveillance based on the technological availability of the new SAR sensor generation on the Cosmo-skymed earth observation constellation.
- Navigation and localisation services based on EGNOS (in future GALILEO) which is the service differentiator with respect to the GPS application, supporting in critical situations better accuracy, availability and

integrity; examples are described for use in critical areas, where accurate localisation and integrated communications are essential.

- Extension of localisation services for indoor applications during emergency activities such as fire events in buildings, allowing accuracy better than 2 meters and limited essential communications toward access satellite points external to the building.
- Last but not least new communications systems, based on transportable Skyplex units, interfaced to a high interoperability module, which enables local communications and can interface WiFi, Wimax and TETRA networks.

Finally an evolutionary scenario for the next 10 years is presented including the evolution of the Earth Observation systems, the integration between EO systems and DRS satellites, the integration of the various positioning satellites towards an integrated GNSS and the new powerful broadband mobile and fixed communication satellite services, which are the space contribution to make the transformation of Homeland Security achievable reality.

There are more things in heaven and earth, Horatio,
Than are dreamt of in your philosophy
(Amleto)

22.2 Foreword

Homeland Protection is the ultimate process, representing the most complex system of systems for human imagination. All the sensitive activities, critical infrastructures, transportations, noticeable sites, green blue and air borders, mass events, each single human life have to be protected against natural and anthropic threats.

The risk of terrorist attacks has to be managed and minimised in a sustainable way, with minimum changes in people way of life, optimising the investments and limiting the presence of invasive infrastructures.

In addition terrorists modify with time their targets and the Homeland Security Process shall prevent if possible, decide the soonest and fast react to any kind of threat, even unsuspected.

Satellite systems are therefore substantial contributors for the intelligence data gathering and elaboration, communications and reaction phases which are the basic elements of the Homeland Security adaptive process.

22.3 Satellite for Security and Safety

Satellite systems can supply the maximum performance, delivering integrated earth observation and localisation/communications services over wide geographical areas, with the lightest possible infrastructure: they are therefore the necessary framework

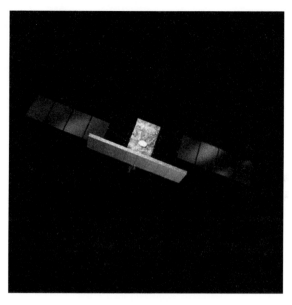

Figure 22.1 Cosmo skymed Artistic View

of any feasible Homeland Security Process, to be integrated in an interoperable scenario with the other local surveillance and communications systems.

The awareness of the importance of the above considerations is well demonstrated by the GMES programme, which is defining an integration strategy between satellite and terrestrial (including air and maritime platforms) systems.

COSMO-Skymed (Figure 22.1) is the newest European observation dual-use constellation with 4 satellites, based on SAR (Synthetic Aperture Radar) sensors.

It will have a repetition revisiting time lesser than 8 hour, a resolution better than 2 meters, no practical limit due to weather conditions and day or night observation time.

The first satellite is in orbit and is delivering imaging data to the receiving earth station.

New communications satellites are going to support more services, based on extensive use of Multi services over IP. Athena Fidus is the new Italy-France communication mission, supplying broadband services for fixed and mobile services dedicated to Security and safety applications.

GALILEO will deliver new localisation and navigation services to improve safety and security. In particular istitutional users will utilise the dedicated Public Regulated Services (PRS) with peculiar performance with respect to robustness, resilience to jamming and cryptography.

Satellite is therefore the key to keep adequate monitoring and reaction.

The Figure 22.2 presents a view of the main risks to face, indicating in a darker blue where satellite technology can supply meaningful support.

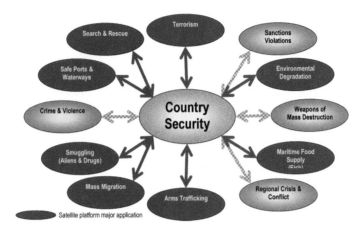

Figure 22.2 Security Risk Scenario

The value added performance are obtained from the integration and fusion of satellite with other technologies,

As it is typical of complex systems the Country security is the governance capability to minimise the overall risk in efficient, fast, adaptive iterative procedure.

Satellite Services cover a lot of security and safety applications for Maritime Safety, such as Search and Rescue Maritime Security, such as Drug Interdiction. General Enforcement of Laws and Treaties Alien Migrant Interdiction Maritime Mobility, such as Lightering Zone Enforcement, Foreign Vessel Inspection, National Defense, such as Homeland Security, General Defense Operations, Maritime Interception Operations, Military Environmental Defense Operations, Port Operations, Security, & Defense Peacetime Military Engagement, Coastal Sea Control, Protection of Natural Resources, such as Marine Pollution Enforcement & Response Living Marine Resource Enforcement.

The main technology families contributing to Homeland Protection are shown in Figure 22.3.

They are articulated in the ones relative to ICT (command, control, computer and communications), the ones dedicated to the sensor technologies, the physical protection technologies, the support system technologies, such as Decision support technologies, which are again ICT, but more addressed to the homeland protection process and the reaction technologies, which are based on the reactions platforms and the systems to drive the reaction itself.

22.4 Added Value Services

Satellites are Key elements to support the scenario envisaged by Homeland Protection: from the point of view of the general Architecture with a high level of Configurability, of robustness, of redundancy and the intrinsic capability to be graceful degradable.

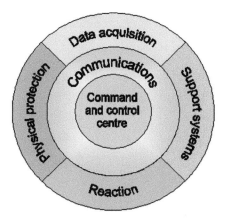

Figure 22.3 ICT Technologies and Security Phases

Figure 22.4 GALILEO Artistic View

For the variety of digital multi media services available in a well integrated form such as:voice, IP service, other Multimedia Services and data transfer Services.

For the typologies of under served by the same platform: Fixed communications, Hand held terminal communications, Mobile vehicular communications, Naval communications, Airborne Communications.

In addition the satellite network allows to overcome the network hierarchy, to cover the territory without any physical infrastructure and all Solutions can be equipped with GPS-EGNOS (in future GALILEO Figure 22.4).

Satellite platforms allow interoperable IP wireless communications (Figure 22.5).

All the communications are monitored by a suitable and simple Control centre, including GIS support to the Crisis Event and 3D Scene analysis, interoperability with external available Data Baseband event simulation for Training and Decision support. Satellite Support the Network Centric Communications vision.

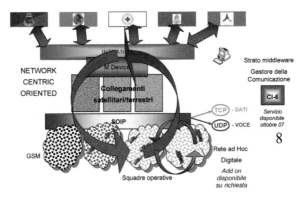

Figure 22.5 Interoperable IP Satellite Platform

The Network Centric operational architecture will closely couple capabilities of Sensors, Decision-makers (command and control), Security enforcement/ engagement unit. Satellite systems are the key element for creating the overlay network among the existing networks.

The importance of the satellite systems as the key element for creating the overlay network among the existing networks.

The satellite control centre has the general characteristics of all the control centres, but the peculiarity is the ease of managing data from sensors, which produce a huge mass of data, the sensors can be optimised to support the monitor functions performance required by the type of risk and reaction, data base of consistent data can be used for forecasting and statistic in a easy and user friendly way, allowing Security Easy configuration upgrading, Built in training, Redundant data base synchronisation, Network statistics and metrics availability.

Satellites support awareness of the activity on field using sensors (SAR and Optical), allow to integrate and support data fusion.

Localisation satellites support on ground sensor georeferencing and navigation of mobile platforms (UAV, Helicopters, airplanes.).

Earth Observation Satellites improve meaningfully Coastal Monitoring, Disaster Monitoring, Law Infringement, Coastal and Environmental Monitoring, Cartography and Mapping, other services, such as civil protection, transport, geology, insurance, hydrological resources.

Satellite data are fundamental for creating Different layers of information data and to use available data on the geographical map superimposed to improve the overall accuracy (Figure 22.6).

A virtual 3D scenario on which it is possible to navigate can be created, starting from available data.

This vision allows a better management during the emergency phases, due to the realism of the information that allows analysis on line of sight for communication, network deployment, etc.

Oil spill is the typical case for which SAR information are usable to monitor (via satellite), controlling during a disaster (via sensor on airplane) and reaction

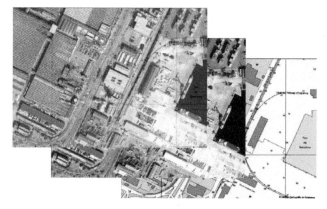

Figure 22.6 Multilayer Image Fusion

Figure 22.7 3D scenario

measuring the spill area extension and positioning the reaction antipollution fences (Figure 22.8). COSMO-skymed will allow a new generation of data and product such as:

- Coherence maps
- Digital Terrain Models (DTM)
- Image co-registration
- Image mosaic
- Image filtering

A nice example of the enriched information is shown in , using optical and SAR images layered to monitor a critical infrastructure (Power Distribution Network) taking advantage of the easy interpretability of optical images combined with the better visualisation of the metallic towers of the Power distribution network.

Figure 22.8 Antipollution fences

Figure 22.9 Optical and SAR images to detect energy infrastructures

22.5 Conclusions

Satellites are going to play a key role in the overall vision of Homeland Security, putting together the essential characteristics of security as an adaptive process.

Communication satellite IPv6 oriented to support the Network Centric Communications Principle, Navigation satellites to support accurate information on location of people and targets, Earth observation satellites to deliver the maximum of the scenario awareness.

In this complex (but not complicated) scenario the possibility of using Data Relay Satellite for coordination and data distribution in real time is an important plus to lever the satellite constellation performance.

Chapter 23
Meteorology in the Dual Use Perspective

Massimo Capaldo

EUMETSAT

23.1 Introduction

Almost 50 years ago, following the Sputnik launch in 1957, a report for the U.S. House Select Committee on Astronautics and Space Exploration predicted that "... great improvement in weather forecasting will become possible with a satellite providing rapid overall data on cloud cover and atmospheric transmissions, albedo...".[1]

In addition and already addressing the dual use of such information the Committee expressed the view that, specifically, cloud-cover photography was considered to be highly beneficial to help intelligence efforts and in support of planning military operations.

Clearly, these predictions were proved to be accurate.

From a purely meteorological point of view, monitoring weather conditions over the globe was greatly improved by the newly acquired knowledge of the current exact location of atmospheric perturbations (figure 23.1) which was earlier estimated by other conventional observations, when available, such as surface winds, pressure tendencies and radio soundings. This specific knowledge allowed a better extrapolation of the local weather conditions as well as the building up of an improved picture of the likely short term evolution of the weather patterns.

Clearly the quality of information has largely improved over the years and equivalent resolutions are now gathered from geostationary satellites which cover larger areas in a single picture (figure 23.2).

23.2 The Operational Continuity

The large benefit acquired by the direct use of remote sensing platforms have motivated , over the years, the strong requirement to ensure the operational continuity of observations from space.

This in the beginning was mainly ensured by US but was followed later by a number of Space Agencies over the world and in Europe by international organizations such as the European Space Agency (ESA) and an European organizations dedicated to the exploitation of meteorological satellites (EUMETSAT).

Both sun synchronous and geostationary satellites have been launched to satisfy the need for comprehensive meteorological information from space in terms of resolution coverage and frequency of observations and roadmaps for the future

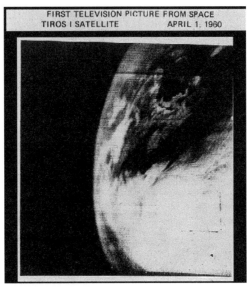

Figure 23.1 The first image from TIROS1 — 1 April 1960 (NOAA archive)

Figure 23.2 Image from meteosat second generation satellite (EUMETSAT archive)

have been drafted and programs started in order to ensure a long standing observation capability from space (figure 23.3 is providing just an example of the EUMETSAT Satellite Programs over the years).

23.3 The Exploitation of Data

The use of data from the various satellite has increased correspondingly to the improvement of the quality of their observations. Together with the direct use of

OVERVIEW OF EUMETSAT PROGRAMMES

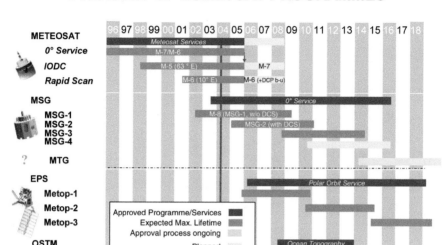

Figure 23.3 EUMETSAT satellite programs

image data , the use of data for global numeric weather predictions, in particular, has been changing from assimilating retrieved temperatures and moisture profile provided by satellite operators to direct assimilation of the measured radiances. The radiance differences between the model first guess estimates of the likely measures and the real measured values can be used to influence the model temperature, water vapour and other useful fields (such as ozone) to obtain an optimal fit between all the observations assimilated and the model first guess (typically a 6 hour forecast). This direct use of radiances has been operational in a number of Numerical Weather Prediction (NWP) Centres since the 90ties and has led to significant improvements in the quality of the NWP analyses and forecasts [2].

This improvement has been particularly striking where forecast are affected by lack of information. As an example radiosonde measurements of atmospheric parameters are very sparse over large areas of the southern hemisphere. Therefore the exploitation of radiation information from satellites have determined a substantial improvement of the quality of predictions, particularly striking over the southern hemisphere where the land masses are less expanded and conventional information is more sparse. Figure 23.4 shows the increase of skill acquired over the years by Numerical Weather Prediction models. When the anomaly correlation of the 500 hpa geopotential field exceed the 60% level, experts consider meteorologically significant the corresponding prediction. From the figure below two things can be observed: first, the overall increase in the quality of predictions over a 25 years time frame but secondly the larger increase of skill over the Southern Hemisphere where satellite information is essential for defining the initial state of the atmosphere [3].

Anomaly correlation of 500hPa height

Anomaly correlation (%) of 500hPa height forecasts

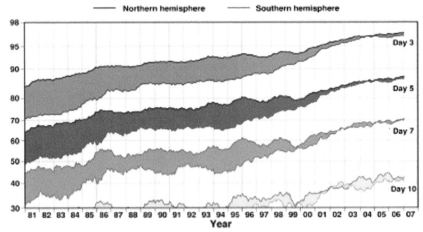

Figure 23.4 Skill improvement 1981–2007 of the NWP model from the european centre of medium range weather forecast (updated version provided by A.Simmons, ECMWF)

Figure 23.5 Image provided by centre de meteorologie spatiale (courtesy of Meteo-France)

However, while space-based observation increasingly involves Low-Earth Orbit satellites (either from operational or from R&D environmental missions) that provide essential input to numerical modelling, the constellation of operational meteorological geostationary satellites remains the backbone of permanent and near-global monitoring of the weather situation.

Infrared composite imagery as shown in figure 23.5 is produced on an operational basis with data from GOES-West and GOES-East (NOAA, USA), Meteosat and Meteosat/IODC (EUMETSAT) and MTSAT (JMA, Japan). Additional

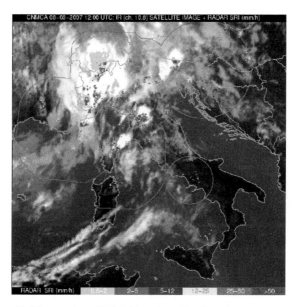

Figure 23.6 Integration of RADAR and satellite data (Courtesy CNMCA)

satellites like the FY-2 series (CMA, China) contribute to strengthen the system and ensure its operational continuity.

The great advantage of geostationary satellites in terms of frequency of observation has allowed the development of specific applications to merge corresponding information gathered from a whole series of remote sensing equipment. The image below shows just an application from the Italian Meteorological Service operated by the Italian Air Force providing a composite image of radar and satellite sensors.

The information can even be expanded to include lightning detecting units (not shown).

Both the resolution characteristic of the current instruments, together with the addition of more and more channels to the observing sensors has allowed the monitoring function to be extended to specific phenomena. Therefore, additional information such as composition of clouds within severe weather systems (such as the hurricane Isabel) as well as monitoring of dust storms has also become available as comparisons of different generations of satellite instruments are showing in in figure 23.7.

In more detail, it can be added that new generation satellites have advanced our ability to identify cloud properties, composition, dynamics and precipitation, by using enhancements in spectral bands, spatial and time resolution.

This allow forecasters to diagnose and nowcast a selection of phenomena, such as:

- **Fog**, by detecting warm layer clouds with small drops.
- **Drizzle**, by identifying clouds with large drops.
- **Rain clouds**, by thick clouds with tops of ice or large drops.

8 September 2003, 12:00 UTC
Hurricane "Isabel"

 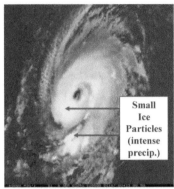

MFG IR Channel i   MSG RGB 05-06,04-09,03-01

23 January 2004, 10:00 UTC
Dust Storm Middle East

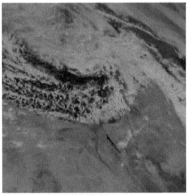 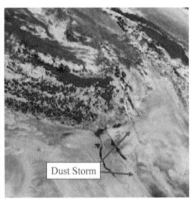

MFG VIS Channel   MSG RGB 01,04r,07-09

Figure 23. 7 Comparison between different resolutions and channels of Meteosat
First generation satellite vs Second Generation for the ISABEL hurricane (upper)
and a dust storm in middle-east (lower)

- **Intense convective storms**, by cold tops with small ice particles, and rate of expansion of the anvils.
- **Dissipating convective storms**, by warming and thinning of the glaciated tops.
- **Multi-layer clouds**, by renewed small particles in the base of each higher layer.
- Supercooled water clouds with **icing aviation hazard**, by small particles of clouds with top temperatures of $0°C<T<-30°C$.
- **Thin clouds**, by large brightness temperature differences.

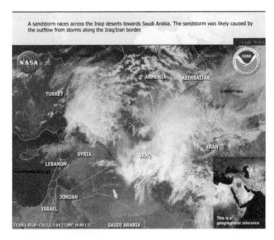

Figure 23.8 Sandstorm in the Arabian peninsula (NOAA)

Figure 23.9 ETNA volcanic ashes trail in Sicily

23.4 The Dual Use

In addition to the different applications mentioned above, sensitive information related to exceptional events has also become available in quasi-real-time.

As a matter of fact information that can be very helpful for planning operations of various nature (support to civilians, military interventions, air traffic management) are of widespread availability in the most recent times.

Useful examples are images of sandstorms or volcanic eruptions which might affect air navigation in nearby airports with significant impact on safety and local economic conditions.

Also some catastrophic events unrelated to environmental issues can be detected and if necessary monitored.

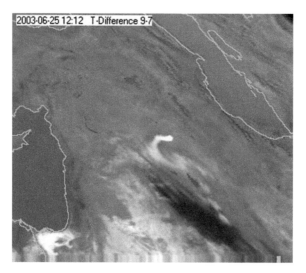

Figure 23.10 Explosion and Fire spreading in Iraq — (Courtesy from EUMETSAT)

Figure 23.10 above shows the effect of an explosion and subsequent fire spreading caused by rocket-propelled grenades fired by unknown persons. The clearly visible toxic smoke affected the cities of Qayyarah, Al Shurah and Makhmur in Iraq.

More in general the applications below are used to provide tactical decision aid in a military environment:

- Sea surface temperature (S&R, ops..) , in figure 23.11
- Multi-spectral analysis of the field of view (res. 1-5 km, 15'-6h time interval)
- Water/Ice phase detection within clouds to help avoiding ice formation (geo and polar)
- Ice/snow/water detection on the land surface (geo and polar)
- Nowcasting for clouds and severe weather phenomena (geostationary)
- Temperature vertical profile for support to ballistics (polar)
- Humidity vertical profiles for anomalous propagation of microwave (polar)

23.5 Conclusions

The remote sensing from satellites has been extensively developed for meteorology in the most recent years according to three major concepts:

- devising applications meant to use the information provided in its original form (i.e radiances) to increase the capability of extracting the embedded content in terms of various parameters of interest;

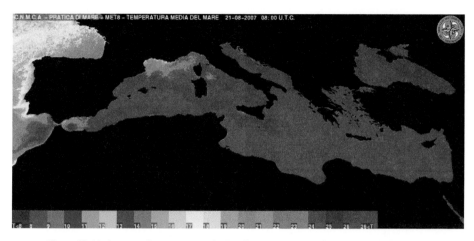

Figure 23.11 A sea surface map over the Mediterranean area (Courtesy from CNMCA)

– taking advantage of the fact that a number of projects have an operational character (i.e. they are the result of multi-lateral commitment to maintain and improve remote sensing capabilities over a long time frame);
– exploiting some specific characteristics of different platforms in terms of area coverage, resolution and accuracy of instruments and most importantly corresponding frequency of repeated observations.

The short review above, without having the ambition to represent a complete framework of the situation, has just sketched some specific remote sensing opportunities offered by current and new generation polar and geo-stationary satellites that allow both direct use of their information and the possibility of applications to be built on their input to feed complex systems that may well be considered of dual use.

References

[1]   U. S. House Select Committee on Astronautics and Space Exploration — *The Next Ten Years in Space: 1959–1969*, — Washington, D.C., 1959.

[2]   Eyre, J.R.- "Progress on Direct Use of Satellite Sounding Radiances in Numerical Weather Prediction." *International Symposium on Assimilation of Observations in Meteorology and Oceanography*, Clermont-Ferrand, France, 9–13 July 1990.

[3]   Simmons, A.J., Hollingsworth, A., "Some Aspects of the Improvement in Skill of Numerical Weather Prediction," 2002, *Quarterly Journal of the Royal Meteorological Society*, Vol. 128, 2002, pp. 647–678.

Chapter 24

Dual Use Applications of High Resolution Radar Imaging

Hugh Griffiths

Defence College of Management and Technology, UK

24.1 Introduction

The techniques of high resolution radar imaging have their origins in radioastronomy, and Ryle and Hewish received the Nobel Prize for Physics for their work on aperture synthesis [13]. At much the same time as this work it had also been realised that the cross-range resolution of a sideways-looking airborne radar (SLAR) could be improved by filtering (a technique known as *Doppler beam-sharpening*) [17]. These ideas were pursued and developed at the Universities of Illinois and Michigan [16]. Since then numerous laboratories and organisations all over the world have built and operated synthetic aperture radar (SAR) systems.

A first comment is that radar is a day-night, all-weather sensor, able to give substantial area coverage. Whilst many Terabytes of image data are readily generated, the challenge is to devise processing algorithms which can extract the required information from this image data [11, 4]. Radar also provides precise measurements of target range, and at the highest range resolutions, information on target range profiles.

It is easy to show that the cross-range (azimuth) resolution of a stripmap-mode SAR is equal to half the along-track dimension of the antenna of the radar sensor. Higher resolution may be obtained in spotlight-mode operation, in which the antenna beam is steered to keep the target scene in view for a longer interval, allowing a longer synthetic aperture to be formed [14]. As with any radar sensor, the range resolution is inversely proportional to the radar bandwidth. In practical terms this means that the highest achievable resolution will be of order half the wavelength of the radar signal (i.e., a few centimetres) [2, 3]. As an example, Figure 24.1 shows a high-resolution image of an airfield, in which details of building and aircraft targets, and their shadows, are evident.

Imaging radar systems have a wide variety of applications, both with aircraft-borne and satellite-borne sensors, and both for geophysical remote sensing and for military surveillance applications, and there are many examples of cross-fertilization of ideas and techniques from one domain to the other. As such, imaging radar is an excellent example of dual-use technology. The remainder of this chapter provides a brief and subjective description of a number of these applications.

Figure 24.1 Example of 3-look X-band aircraft-borne SAR image yielding 10 cm resolution (after Cantalloube and Dubois-Fernandez [3])

24.2 Military Applications

Important military capabilities that imaging radar may provide include area surveillance, identification and location of hostile targets, and battle damage assessment. Specific techniques of current interest include Ground Moving Target Indication (GMTI), Automatic Target Recognition (ATR), Foliage Penetration (FOPEN), the use of Unmanned Air Vehicles (UAVs), and bistatic operation.

Synthetic Aperture Radar is able to distinguish moving targets from static ones, via the pulse-to-pulse change in phase of the target echo (Doppler shift) as the target moves. GMTI processing consists of coherently subtracting echoes from successive pulses received at antenna subarrays displaced along-track, so that echoes from stationary targets (background clutter) cancel, whilst those from moving targets do not [6]. This technique provides a powerful means of detecting moving vehicles, which is exploited in GMTI modes of systems such as Joint STARS and ASTOR.

The ability automatically to classify and recognise targets in radar imagery represents something of a Holy Grail, and whilst it has been the subject of considerable research efforts in many countries over the past decade, it remains a difficult problem that cannot yet be said to be solved. Fundamentally, it depends on the generation of high-resolution imagery with adequate signal-to-noise ratio and free from artefacts. Consideration of the equivalent problem of visual recognition of objects shows that humans exploit a variety of information: target size, shape (both of the object and of its shadow), colour, as well as the context in which a given target may be expected to be seen [1]. A human can also choose to look at an object from a different perspective to provide additional information, and similar approaches have been shown to be of value in radar target recognition [15]. In radar imagery, techniques such as interferometry and polarimetry can also provide additional information.

In a military context it is also necessary to be aware of the techniques of Camouflage, Concealment and Deception (CC&D), so adversaries may modify the signatures of genuine targets or deploy decoy targets, in an attempt to confuse our knowledge and decision-making.

At radar frequencies in the VHF or UHF bands, the penetration through foliage canopies can be significant, allowing detection of targets hidden in forests. In addition, such frequencies may be effective against stealth targets whose physical dimensions will cause resonance effects. However, a SAR at these frequencies poses some particular design challenges: (i) since the carrier frequency is low, the fractional bandwidth necessary to achieve adequate range resolution is substantial, possibly exceeding 100%; (ii) thus the radar hardware and (particularly) the antenna need to be able to handle this high fractional bandwidth; (iii) since the antenna beamwidth will be broad, the SAR processing has to cope with significant range migration; (iv) there are already many broadcast and communications users in these bands, so special techniques are needed to avoid mutual interference.

Despite these challenges, interest in FOPEN SAR has been substantial, and several experimental systems have been built and evaluated. Notable among these is the CARABAS system, developed over the past two decades by the Swedish Defence Research Agency, FOI [9]. This has required substantial innovative research, as well as investment by funding agencies. The most recent version of the CARABAS system uses a bandwidth extending over 20–90 MHz. More recently, the LORA system uses 220–420 MHz, allowing higher resolution.

UAVs find increasing use in military applications for a variety of reasons, amongst which are that they can operate in scenarios which are hostile to humans, and that they can be smaller, with significantly lower signatures, than manned platforms. They range from large, long-endurance platforms such as GLOBAL HAWK, to micro UAVs similar to those used by model aircraft enthusiasts. Such platforms may be used to carry SAR sensors. Important considerations are the weight and power consumption of the sensors and of the telemetry used to deliver the data to the ground, and several organizations have developed lightweight low-power SAR payloads for UAVs.

Bistatic configurations have some significant attractions, both because the power consumption associated with the transmitter can be avoided, and because the bistatic geometry may offer improved detection performance against low-signature targets [8, 18]. In addition, many of the problems of positioning, motion compensation and synchronization which have proved difficult in the past are now solvable with GPS. More generally, it can be seen that there are several configurations, with moving transmitter, moving receiver, or both, and with aircraft, satellite or UAV platforms.

24.3 Civil Applications

The first spaceborne SAR system was carried by NASA's SEASAT satellite in 1978. This only lasted for 3 months, until a power supply fault cut short its life. Nevertheless, it provided a wealth of data, and demonstrated the value of spaceborne SAR for a wide variety of applications in environmental monitoring. Subsequently NASA, the European Space Agency, Japan, Canada and several other Agencies have built and flown satellite SAR systems of increasing sophistication, now often with multiple frequency bands and polarimetric capability. Other current techniques of interest include interferometry and differential interferometry.

SEASAT was designed for the remote sensing of the oceans, and as well as its SAR also carried other instruments, providing information on ocean surface elevation, wave height, wind speed and surface temperature. However, its orbit pattern also took it over land and polar regions, and provided a great deal of data from these types of surface as well. Satellite-borne radar is particularly suitable for polar remote sensing, both because such regions are remote, inhospitable and dark for much of the year, and because they are of great importance in the study of the effects of climate change.

Since the radar scattering from targets may depend on frequency and signal polarization, multi-band and polarimetric operation can yield important additional information. In particular, the polarimetric scattering properties of dihedral and trihedral features that may be characteristic of man-made targets provide useful information in distinguishing man-made from natural targets [5].

The techniques of SAR interferometry date back originally to the 1970s [7]. Essentially, if two SAR images of a target scene are obtained from slightly-displaced receive antennas, and co-registered, the phase differences between corresponding image pixels yield an *interferogram*, in which the phase differences are a function of a number of quantities including the height of the target within that pixel. If the other parameters are known, or can be estimated, the target height can be derived from the phase difference measurement, over the whole SAR image, yielding a 3-D representation of the target scene to the same resolution as the original SAR images. This allows the derivation of 3-D maps of the Earth's surface, and in high-resolution imagery can give information on target shape.

A further development is that of differential interferometry. Here, two interferograms of a given target scene are obtained, separated by some interval of time, and subtracted one from the other. If the target scene has remained constant, the result will be a constant phase difference, but if there has been any vertical movement in the intervening time, this will give interference fringes due to the change in elevation. This is an extraordinarily sensitive way of detecting elevation changes. This technique may also be used in high-resolution SAR imagery to provide a means of coherent change detection. Thus if a set of targets are present in one scene but not in another, the differential interferometry will not only show the targets, but also give information on their sizes and shapes [10].

24.4 Where Next?

This brief overview has attempted to give a glimpse of some of the techniques of radar imaging for both military and civil applications. It should be apparent that there is strong synergy between the two, and techniques developed in one context can find application in the other. The keys to success in all of these applications are firstly the ability to produce high-quality, artefact-free imagery, and secondly to be able to extract the wanted information. We can also note particular interest in bistatic SAR and the use of UAVs.

The conclusion to be drawn is that the techniques of Synthetic Aperture Radar are in vibrant health, and that new ideas and refinements of old ones, for all sorts of applications, can confidently be expected in the coming years.

Acknowledgements

I acknowledge invaluable discussions with many people from whom I have learned much about synthetic aperture radar over the years. I would particularly like to mention Chris Baker, Chris Oliver and Richard White. I am also grateful to René Garello, Anders Gustavsson, Lena Klasén, Björn Larsson, Pierluigi Mancini, Jone Sæbbø, Lars Ulander, Michele Vespe, Aric Whitewood and Gill Yates.

References

[1] Blacknell, D., "Contextual information in SAR target detection", *IEE Proceedings Radar, Sonar and Navigation*, Vol. 148, No. 1, February 2001, pp. 41–47.

[2] Brenner, A.R. and Ender, J., "Airborne SAR Imaging with Subdecimetre Resolution", *Proceedings EUSAR 2004 Conference*, pp. 267–270.

[3] Cantalloube, H. and Dubois-Fernandez, P. "Airborne X-band SAR Imaging with 10 cm Resolution — Technical Challenge and Preliminary Results", *Proceedings EUSAR 2004 Conference*, pp. 271–274.

[4] Capraro, G., Farina, A., Griffiths, H.D. and Wicks, M.C., "Knowledge-Based Radar Signal and Data Processing: A Tutorial Introduction", Special Issue of *IEEE Signal Processing Magazine* on Knowledge Based Systems for Adaptive Radar Detection, Tracking and Classification, Vol. 23, No. 1, January 2006, pp. 18–29.

[5] Chaney, R.D., Burl, M.C. and Novak, L.M., "On the Performance of Polarimetric Target Detection Algorithms", *IEEE International Radar Conference RADAR'90*.

[6] Ender, J., "Detection and Estimation of Moving Target Signals by Multi-Channel SAR", *AEU International Journal Electronic Communication*, 50, No. 2, 1996, pp. 150–156.

[7] Graham, L.C., "Synthetic Interferometric Radar for Topographic Mapping", *Proceedigns IEEE*, Vol. 62, No. 6, 1974, pp. 763–768.

[8] Griffiths, H.D., "From a Different Perspective: Principles, Practice and Potential of Bistatic Radar", *Proceedings International Conference RADAR 2003*, Adelaide, Australia, September 2003, pp. 1–7, 3–5.

[9] Larsson, B., "The CARABAS-II VHF SAR system", invited keynote address, European Radar Conference *EuRAD*, Manchester, September 13–14, 2006.

[10] Massonet, D., Rossi, M., Carmona, C., Adragna, F., Peltzer, G., Feigl, K. and Rabaute, T., "The Displacement Field of the Landers Earthquake Mapped by Radar Interferometry", *Nature*, Vol. 364, 8 July 1993, pp. 138–142.

[11] Oliver, C.J. and Quegan, S., *Understanding SAR Images*, Artech House, 1998.

[12] Rihaczek, A.W., *Principles of High-Resolution Radar*, McGraw-Hill, 1969; reprinted by Artech House, 1996.

[13] Ryle, M., "A New Interferometer and Its Application to the Observation of Weak Radio Stars", *Proceedings Royal Society A.*, Vol. 211, March 1952, pp. 351–375.

[14] Ulander, L.H., "A New Equation for SAR Spatial Resolution", *Proceedings EUSAR'96 Conference*, Königswinter, pp. 389–392.

[15] Vespe, M., Baker, C.J. and Griffiths, H.D., "Radar Target Classification Using Multiple Perspectives", *IET Radar, Sonar and Navigation*, Vol. 1, No. 4, August 2007, pp. 300–307.

[16] Wiley, C.A., "Pulse Doppler Radar Methods and Apparatus", US Patent No. 3, 196, 436, 1954.

[17] Wiley, C.A., "Synthetic Aperture Radars — A Paradigm for Technology Evolution", *IEEE Transactions Aerospace & Electronic Systems*, Vol. AES-21, pp. 440–443, 1985.

[18] Willis, N.J. and Griffiths, H.D. (eds), *Advances in Bistatic Radar*, SciTech Publishing Inc., Raleigh, NC, ISBN 1891121480, 2007.

PART 4

Industry Outlook on Dual Use

Chapter 25

A Contribution From Rheinmetall Italia SpA

Vittorio Dainelli

Rheinmetall Italia S.p.A.

On July 31st 2007, Oerlikon Contraves S.p.A. (OCI) changed its name to Rheinmetall Italia S.p.A.

What Will Not Change	What Will Change
– the same experienced, flexible and reliable team – the same established competencies in advanced RF technologies and applications – the same long-lasting reputation and acknowledged leadership in the management of complex systems – the same tradition of orientation to the customer and problem solving capability	– All strengths of the Rheinmetall group in one hand – one interface only for all the products of the Rheinmetall Group – system of systems capability – extended range of competencies and technologies – global presence

RhI envisaged on Microelectronics for RF unit the most strategic line of research to be follow in order to have a common base for the development of new "fully European" products for future dual use applications (defence and security).

A technology road map has been identified and internally financed for developing and manufacturing a strategic set of "core chips" on GaAs (MMIC's) or on Si (ASIC's) that can be used in modular way for the realisation of subsystems in a range of frequency from digital to MM wave with a fully Italian/European solution.

The advantages offered in term of weight, size, power efficiency, reliability, flexibility, repeatability, cost, performance, combined with the extremely wide frequency coverage (from Digital to mm Waves), make MMIC and ASIC developed by Rheinmetall Italia S.p.A. the ideal solution for many military, civil and space applications.

Thanks to their modularity, they can find application for:

- **L, S, C, X, Ku, Ka, W band (1 to 95 GHz) radar sensors**
  - MILITARY
    - AIR Defence (Search and tracking)

- ■ Ground based
  - □ Stationary
  - □ Mobile
- ■ Naval
- o C-RAM

X Band 3D 25 / 35 / 50 km Range

Tracking and Fire Direction X / Ku Band Naval

- • CIVIL
  - o Surveillance and Security
  - o Traffic Control
    - ■ Road
    - ■ Railway
    - ■ Vessels
    - ■ Airport
    - ■ FOD (Fallen Object Detection)

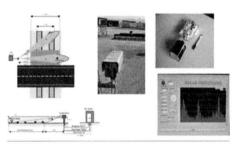

Limited Area Foreign Object Debris WARNING RADAR (W band)

W Band Smart High Resolution radar MK2

- • SPACE
  - o W band High definition 3D Imaging (Lunar surface Survey and Mapping)
  - o W Band Advanced Radar for Debris Early Notification (WARDEN)

- **Broad Band, high data rate and secure communication for space applications**
  - W band InterSatellite Links (ISL)
    - o W band Ground to satellite and satellite to ground communications (WAVE)

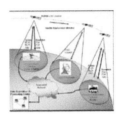

- **multifunctional units for satellites, space crafts and rovers**
    - MM wave payload for the support of rover operations in extraterrestrial environments

In the frame of this contest RhI have developed with the support of scientific institutions in Rome (Tor Vergata University, La Sapienza University) and of European GaAs or Si foundries the following:

- MMIC (Monolithic Microwave Int. Circ.) multifunction modules
  - Tx/Rx modules
  - Integrated Front End units
  - Integrated Exciter units

In particular:

– **X BAND RECEIVER MODULES FOR RADAR SENSORS**

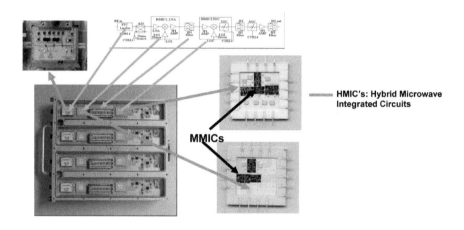

– **MMIC CHIP SET FOR X BAND TX/RX MODULES**

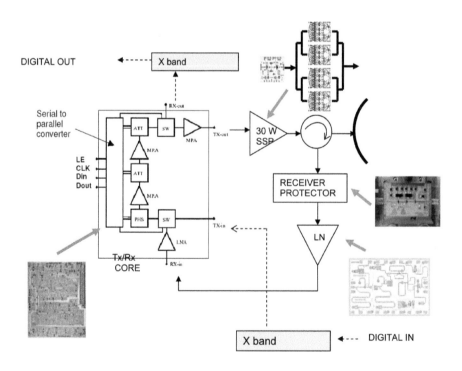

– MMIC'S FOR PERSONALIZATION OF RX AND TX UP TO 95 GHZ AND W BAND FMCW FRONT END

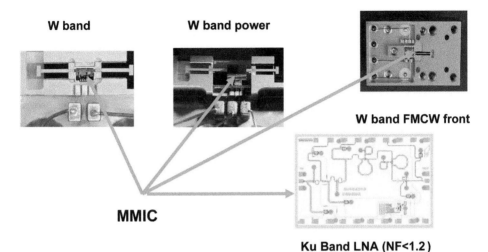

W band

W band power

W band FMCW front

MMIC

Ku Band LNA (NF<1.2)

– ASIC (Application Specific Integrated Circuit)
  • Signal conversion RF to Digital
  • Integrated, customised Digital Signal Processing

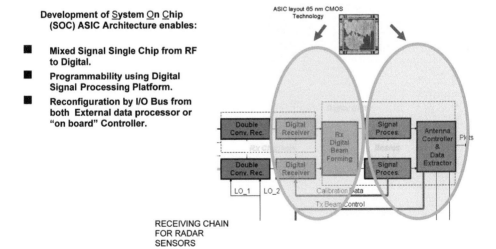

Development of System On Chip (SOC) ASIC Architecture enables:

■ Mixed Signal Single Chip from RF to Digital.

■ Programmability using Digital Signal Processing Platform.

■ Reconfiguration by I/O Bus from both External data processor or "on board" Controller.

ASIC layout 65 nm CMOS Technology

RECEIVING CHAIN FOR RADAR SENSORS

Chapter 26
An Overview of the GALILEO Program

Cosimo La Rocca, Alexander Mager, Tiziano Sassorossi

European Satellite Navigation Industries S.p.A., Rome, Italy

26.1 GALILEO System

The GALILEO System is an independent, global, European-controlled, satellite-based navigation system. Its constellation of satellites is complemented with a Ground Segment providing system and satellite monitoring and control, orbitography, time reference and integrity data.

GALILEO will be the first example of a navigation system completely under civilian control.

This system will be a cornerstone of the Europe high-technologies in the next years.

The implementation phase of GALILEO Program started at year-end 2004, with the so called CDE1 phase.

It required a long and intense preliminary work, preparing the political environment, the industrial configuration and the overall technical specifications.

GALILEO follows an opposite approach compared to the two existing satellite-based radio navigation systems GPS and GLONASS.

Both existing systems have been designed around military needs, and only afterwards the civil market discovered the huge benefits of satellite-based navigation. On the contrary, GALILEO starts from the civil market needs, although opening evident perspectives in terms of dual-use applications.

The GALILEO Space Segment (SSEG) will comprise a constellation of twenty seven operational satellites in Medium-Earth Orbit (MEO), so that at least ten satellites will normally be visible from any point on the Earth's surface, plus nominally three non-operational in-orbit spare satellites, one in each orbit plane. Each satellite will broadcast precise ranging and time signals on four carriers, together with clock synchronization, orbit ephemeris, integrity and other data. A user equipped with a suitable receiver will be able to determine his position with an accuracy of a few meters when receiving signals from just four GALILEO satellites.

The GALILEO Ground Segment is split into two Segments:

- the GALILEO Control Segment (GCS) will control the whole GALILEO constellation and monitor the satellites health,
- the GALILEO Mission Segment (GMS) will produce navigation data and up-load them to the satellites for subsequent broadcast to users. The key

elements of the data, i.e. clock synchronization, orbit ephemeris and integrity, will be calculated from measurements made by a worldwide network of stations.

The GALILEO System will also provide an interface to Service Providers. The Service Providers will give the users a point-of-contact to the GALILEO system, will provide a variety of value-added services and will play a role in collecting fees.

This interface may also include provision of specialist data, such as clock and ephemeris history and predictions to specialist scientific users.

26.2 GALILEO Main Features

26.2.1 GALILEO Security Architecture

The security aspect in GALILEO has two main objectives

- Protection of the entire system
- Provision of a secure Public Regulated Service (PRS).

The safety and security requirements cover all elements and levels of the GALILEO system, but, contrary to a military system like GPS or GLONASS, GALILEO will be both certifiable and accreditable.

The implementation of security in GALILEO led to the use of "military" technology e.g. key management and encryption codes for PRS.

The satellite design does not meet military standards (no "military hardening") whereas the ground infrastructure (control centres, ULS/TT&C/Sensor stations and network) is well protected e.g. against terrorist attacks.

The GALILEO System architecture, with specific reference to security aspects, is presented is Figure 26.1, where arrows in different colors show the various data links with their own crypto levels.

26.2.2 GALILEO Services: The Public Regulated Services

The GALILEO System will be able to deliver a signal in space that will support the following main services:

- Open Services (OS)
- Safety-of-Life Services (SoL)
- Commercial Services (CS)
- Public Regulated Service (PRS)
- Search and Rescue (SAR)
- External Regional Integrity (ERIS)

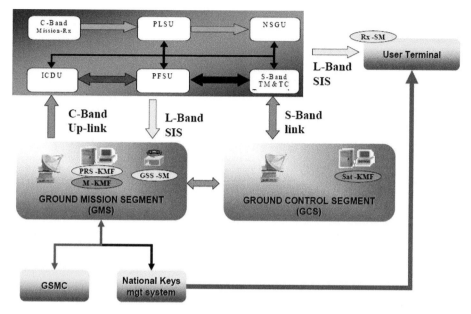

Figure 26.1 GALILEO Security Architecture

GALILEO has been designed around the needs of the civil and commercial market with the potential to be used as well by armed forces. Namely, GALILEO's Public Regulated Service (PRS) is the service of interest for armed forces and with potential military applications.

It provides a robust and access-controlled service for government applications (protected against jamming and spoofing), it is operational at all times and in all circumstances, notably during periods of crisis ("continuity of service") and it is separated from other services that can be denied without affecting PRS operations.

Finally it offers an integrity signal, a feature not available in today's GPS, but planned in the GPS upgrades (M-Code).

26.3 Dual Use Applications

26.3.1 GALILEO as a Dual Use System

Already back in the 1980's Europe decided to develop its own satellite-based radio navigation system for a simple reason: to foster Europe's political, technological and economic independence.

In contrast to the existing US American (GPS) and Russian (GLONASS) systems developed during the "cold war" for the armed forces and consequently operated by military personnel, Europe decided to launch GALILEO as a civil system under civil control with the primary objective to boost commercial applications. However, Europe identified from the very beginning the need and benefit of such system also for public authorities: This requirement led to the design and the development of the so called "Public Regulated Service" (PRS), which is described in the section above.

Regarding military requirements two criteria are still missing though: a "military hardened" GALILEO infrastructure and a clear policy for using PRS.

26.3.2 Exploitation of GALILEO Dual Use Capabilities : Pro's and Con's

Undoubtedly, GALILEO will be perceived and used from the very first day of its operation as a "dual use" product just as telecommunication or earth observation satellites. A further implementation of the "dual use" concept for GALILEO implies a number of advantages and inconveniences.

Advantages:
A military use of the GALILEO system will certainly contribute to Europe's political independence since it significantly enhances the credibility and capability of its military means and tools — e.g. in combination with GMES (Global Monitoring of Environment and Security) — in the context of a converging European foreign policy. Also NATO could very much benefit from a second redundant navigation signal: combining the GPS M code with GALILEO's PRS would clearly increase efficiency and safety of operations, especially in unknown and difficult terrain. Another important aspect is the new generation of SAASM (Selective Availability Anti-Spoofing Module) receivers in combination with the "over the air rekeying" (OTAR) technology implemented in the upgraded GPS IIRM, F and III systems. This new generation of military navigation receivers allows enabling/disabling each single delivered device via OTAR. Another major argument is the growing importance of navigation components integrated in all kind of aerospace and defence products. It is of utmost importance that European industry safeguards its autonomy and deriving global export chances by developing platforms and systems offering a "military" navigation signal under European control.

Inconveniences:
A truly "dual use" product is per definition more complex than a civil one and is consequently more expensive since it has to cope with military requirements. The "military hardening" of products implies several levels of data classification, furthermore a more complex concept for "Command and Control" has to be developed, in order to properly handle military priorities and to segregate the military information, and generally speaking, consumer manufacturers will find it less easy to get access to data due to tightened security requirements.

In many cases the old concept of "dual use" as it has been defined in the past — civil vs. military applications — cannot be defended any longer for several reasons: Civil users like police forces or border patrol units in charge of "homeland security" increasingly require the same type of equipment than their military counterparts due the new threats being for example international terrorism or regions in which asymmetric wars are taking place. Moreover, civil applications such as crisis management, law enforcement or the tracking of dangerous goods are often carried out by personnel and organisations belonging to the Ministry of Defence (e.g. the Italian "Carabinieri" or the French "Gendarmerie"). These examples clearly show that cataloguing systems in civil vs. military ones becomes more and more difficult because formerly called "civil" systems resemble more and more "military" ones. The GALILEO PRS is a perfect example for this new "dual use" concept.

26.4 Conclusions

The GALILEO System is a civil system under civil control well suited and capable of offering one of its services, the PRS, to a military customer.

However the concept of "dual use" is fading away, civil users like police forces or border patrol units in charge of "homeland security" increasingly require the same type of equipment than their military counterparts.

In addition civil applications such as crisis management, law enforcement, tracking of dangerous goods are often involving military corps.

The Public Regulated Service (PRS) is a perfect example for this new concept; however a clear policy for using PRS is not available yet and is urgently needed.

Starting form the current configuration, the GALILEO infrastructure can be further developed in a stepped approach, such that it will be possible in the future to meet also operational needs and requirements of armed forces ("military hardening").

The following key issues at stake are to be considered and solved to progress smoothly in the development of GALILEO Security concept:

- Political agreement of National Governments on concept of use
- Security management concept to be further developed
- PRS accreditation process.

Chapter 27

Dual-Use Technologies & Applications for Space Systems

S. Arenaccio, G. Chiassarini, R. Loforti, F. Petrosino, A. Vernucci

Space Engineering S.p.A., Rome, Italy

Abstract. This chapter dwells on some of the innovative developments performed by Space Engineering S.p.A., an Italian medium-size company active since nearly twenty years in the field of space technologies and applications. Said developments, though conceived and pushed through in civilian contexts, are directly compatible with or could anyway be adapted for their exploitation in the defence world. The overview herein given encompasses ground-segment equipment allowing access to the satellite, for both communications and telemetry/telecommand purposes, as well as networks intended to support innovative applications.

27.1 Introduction

Small-medium enterprises are gaining more and more important roles in the development of novel space systems and applications, also thanks to the very significant software technology advances which took place in the last fifteen to twenty years. On the one hand the development of software-based products is greatly facilitated by the widespread availability of low-cost Personal Computers and the relevant software tools, on the other hand hardware-based products can today largely rely on software-programmable devices (Field Programmable Gate Arrays) and solutions (Software Defined Radio) that can be afforded even by the small players, with no need for great capital investment.

The Rome-based Space Engineering is a good example of a medium-size company which was able to take advantage of the new software technology paradigm for producing equipment and applications having high innovation contents. Such achievements were possible also thanks to the company participation in research program funded mainly by the Italian Space Agency (ASI), the European Space Agency (ESA) and the European Commission (EC).

Among such achievements we here shortly describe some equipment and applications that are expected to also be of potential interest for the defence world. In particular:

- Ground processors for wideband communications (see Sect. 27.2);
- Advanced Telemetry and Telecommand (TT&C) modems (see Sect. 27.3);
- Tracking terminals for communications on the move (see Sect. 27.4);
- Telemedicine network (see Sect. 27.5).

27.2 Ground Processors for Wideband Communications

With the advent of satellite TV and, subsequently, of via-satellite Internet and IP-based applications in general, a growing interest for wideband satellite communications is being paid in the civilian world.

Such tendency is however also being noted in the context of military satellites, which are evolving from systems capable of only supporting voice and narrow-band data delivery toward systems capable of serving much richer applications having high graphical contents.

The broadcast satellite systems, as well as the interactive ones, have traditionally been based upon the Digital Video Broadcasting — Satellite (DVB-S) technology, which allows high-rate data delivery from the Hubs to the User Terminals (Forward-Link). In the Return-Link, i.e. from the user Terminals to the Hubs, the DVB-RCS (DVB-Return Channel via Satellite) standard is commonly being used when such link is implemented via satellite (opposite to systems in which the Return-Link is realized by terrestrial means).

As far as the Forward-Link is concerned, technology advances have led to the recent issue of a second-generation DVB-S standard, called DVB-S2, which offers better performances in the broadcast mode (mainly thanks to the adoption of a more efficient coding solution called Low Density Parity Check code — LDPC) and much better performances in the interactive mode (mainly thanks to the use of an Adaptive Modulation and Coding — ACM scheme which adapts, in real time, the transmission parameters to the current rain-induced attenuation affecting the link).

Space Engineering looks with great interest to DVB-S2, it representing an important opportunity to take-off in the development of wide-scale commercial products. The company is currently completing the implementation of a comprehensive hardware-based real-time DVB-S2 Test Bed in the context of *DEDICATION*, an ESA ARTES-4 co-funded project. Such Test Bed comprises the processing sections of both the Hub and the User Terminals, namely:

- the *Hub Transmitter*, including an Upper-Layer section (accepting IP-packets and encapsulating them while supporting two different Quality-of-Service classes) and a Physical-Layer section (performing ACM, LDPC encoding and modulation with programmable pre-compensation of the satellite / ground station power amplifier non-linearity);
- the *User Terminal receiver*, including a Physical-Layer section (performing ACM demodulation and LDPC decoding) and an Upper-Layer section (de-capsulating IP-packets and outputting them to the exterior);
- the *Test Facility*, including the Hub Server integrated with the Traffic Generator (producing the "wanted" traffic and emulating the presence of other system users), the User Terminal Client integrated with the Traffic Analyser (performing traffic statistics and determining packet loss rates, delay and jitter), the Satellite Channel Emulator (simulating linear and non-linear channel distortions and effects, i.e. interference, frequency error, phase- and thermal-noise, fading), and the overall Test Bed Controller, in charge of configuring all the Test Bed elements and supporting a Graphical User Interface.

Such Test Bed represents a rather rich implementation of the DVB-S2 standard, in that it supports virtually all the meaningful modulation / coding pairs specified by the standard, a very high symbol rate (up to 45 Msym/s) that permits operation in conjunction with the modern wideband satellite transponders, and the so-called Generic Stream IP-packets encapsulation strategy in addition to the more common MPEG-Transport Stream strategy.

Space Engineering is now looking into the possibility of porting the above developments on custom integrated circuits, such devices being indispensable for realizing the processing subsystem of relatively low-cost User Terminals. It should be noted that such terminals would also be compatible with the requirements of *Athena-Fidus*, the new dual-use satellite system that is being realized in cooperation between Italy and France.

For what concern the Return-Link, i.e. from the User Terminals to the Hubs, Space Engineering is currently engaged in an ESA-funded ARTES-5 project (called *AMPIST*) in the context of which the above described *DEDICATION* Test Bed will be complemented by extra equipment making the Test Bed representative of the processing subsystem of a complete two-way satellite system (i.e. Forward-Link + Return-Link). Such enhanced Test Bed incorporates numerous novelties with respect to the present DVB-RCS standard, that will permit to significantly improve its transmission performance, namely:

- an enhanced Forward Error Correction scheme with improved performance with respect to the current DVB-RCS Turbo code;
- a novel framing and layer-2 encapsulation scheme optimized for the adaptive physical layer and matching traffic characteristics;
- high-order modulation schemes (i.e. 8PSK and 16APSK) with ACM and a variety of code rates;
- optimized contention-based access, also in conjunction with Continuous-Phase Modulation;
- interference cancellation techniques intended to suppress co-channel and adjacent channel interference.

With the completion of the *AMPIST* project, Space Engineering will be in the position to exploit opportunities occurring in the context of wideband communications systems that utilize satellite resources in both the Forward- and the Return-Link.

27.3 Advanced TT&C Modems

Space Engineering has been involved in the TT&C business since 1999, when it produced the TT&C modem for the first Italian military satellite *SICRAL*. That development was possible thanks to the experience in the implementation of complex digital hardware, including spread spectrum Code-Division and Time-Division Multiple Access (CDMA and TDMA), gained through the participation in European civilian research projects. The *SICRAL* TT&C modem is of the Direct-Sequence (DS) Spread-Spectrum (SS) type; it interfaces with the ground station subsystems at 70 MHz, with transmission to / from the satellite in the EHF frequency range. The *SICRAL* modem is today still in operation with no failures being recorded across the 7-year operational period.

Subsequently in year 2000, in the frame of a comprehensive real-time military satellite telecommunication system emulator, Space Engineering developed a TT&C Test Bed for the Korean *MILSATCOM* system, in the context of which both the on-ground and the on-board TT&C terminals were implemented, adopting a Frequency-Hopping technique.

In year 2004, Space Engineering was put in charge of developing the TT&C modem for the *KOREASAT 5* military satellite telecommunication system of the Korean Defence Administration. In addition to the proper SS TT&C modem, the developed equipment incorporated a ranging processor based on a loopback signal.

In year 2005 / 2006, Space Engineering developed the DS-SS TT&C modem for the Italian military satellite *SICRAL-1B*.

Finally, in year 2006 / 2007, Space Engineering produced a modem & ranging processor for the TT&C Satellite Check-Out Equipment of *GALILEO*, the forecoming European radio-localization system. Such equipment supports both the standard and the SS TT&C modes. This last mode performs the following main functions:

- provides anti-jamming capability to the telecommand (uplink) and telemetry (downlink) channel by applying a pseudo-noise data spreading technique;
- allows ranging measurements, by providing the transmit and receive code time delay.

In addition to the SS ranging, the developed equipment also supports multi-tone ranging as well as code ranging.

As a recognition of its remarkable experience in the TT&C field, ESA recently awarded Space Engineering, on competitive basis, a research project dealing with future multipurpose TT&C systems & techniques. The main rationale behind that activity is to investigate new TT&C communication systems and techniques that, leveraging on the recent remarkable advances in signal processing techniques and technologies, can serve a broad range of possible missions (e.g. from near earth to deep space), in the perspective of avoiding the use of expensive ad-hoc solutions to cope with the different mission requirements.

27.4 Tracking Terminals for Communications on the Move

One of the areas of specific Space Engineering interest is the implementation of terminals for wideband communications from ships, aircrafts, trains and vehicles. It should be noted that wideband communications can only be implemented, at reasonable cost, in the Ku and Ka-band ranges, while L- or S-band systems (e.g. Inmarsat) are more suitable for narrowband communications.

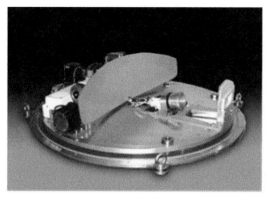

Figure 27.1

Figure 27.2

In this section a terminal for high-speed trains applications is addressed, with regard to which the main Space Engineering responsibility was in the low-profile tracking antenna (actually designed and built by TeS, a 100%-owned Space Engineering subsidiary, on contract to Eutelsat). For that purpose an ad-hoc K_u-Band antenna system had to be designed. The basic idea was to realize a dual multifocus shaped reflector as compact as possible so as to fit the specified roof-top train envelope while keeping simple mechanical horizontal and vertical antenna scan capabilities as well as very low RF chain losses.

The antenna optics is an offset dual reflector antenna having foci on the main planes.

The leading reason behind this choice is related to the need of having a low profile antenna, to be compliant with the train envelope. High-speed train gauges impose huge constraints on the antenna profiles because of the presence of tunnels along the railways combined with more general aerodynamic requirements. Moreover the antenna is capable of fully mechanical horizontal and vertical scans: vertical beam steering is achieved only rotating the main dish thus keeping fixed the remaining part of the antenna assembly while the horizontal antenna pointing is performed by rotating the platform where the antenna is placed.

The antenna system design aims at optimizing the available room in order to achieve the maximum antenna radiating energy, thus avoiding any blockage/scattering coming from the structural parts.

The antenna RF performance has been optimized in order to get a very good antenna gain at both the receive and transmit frequency bands, thus limiting the high-power power amplifier cost. Reflector shaping techniques have been used to achieve the maximum antenna gain (including a A-type sandwich radome) within the whole frequency & scanning ranges.

In order to qualify the low-profile antenna several field trials have been carried out on-board high speed trains throughout the Europe, consistently with the very severe railway gauges. Encouraging results have been achieved, both in terms of the overall achieved performance and from the system reliability point of view.

Same pictures taken during the test campaigns, lasting more than one year, are shown in the following figures 27.3-27.8.

Figure 27.3

Figure 27.4

Figure 27.5

Figure 27.6

27.5 Telemedicine Network

Generally speaking, telemedicine networks can be used in support of several types of medical activities, e.g. monitoring, control of medical equipment, home care, first aid, consultation, access to medical files, sharing medical information, etc. Telemedicine services are considered to be among the most efficient and cost-effective solutions to improve quality of health care for patients located in remote sites. In fact, through the utilization of an efficient telecommunication infrastructure a great deal of data can be shared among a number of medical personnel and entities, allowing faster diagnosis, avoiding travelling and patient transportation in various situations.

The importance that telemedicine can have in military contexts is self-evident and does not deserve further consideration.

Thanks to their characteristics, satellite systems are very suitable to constitute the core infrastructure of telemedicine networks; as a matter of fact satellites can

Figure 27.7

Figure 27.8

provide broadband access ubiquitously over very large areas, including remote or impervious locations where terrestrial telecommunication infrastructures are often not present or are economically not viable. In addition, satellite systems can ensure long-range mobility and are particularly suitable to provide service during emergency when no other networks are unavailable. Satellite systems can be complemented by wireless systems (e.g. WiFi, WiMax) so as to improve service capillarity and ubiquitousness.

In that perspective ASI has co-funded a project called *TELESAL* aiming to set-up an experimental telemedicine service platform by developing and experimenting new applications able to support various telemedicine services (tele-monitoring, tele-diagnosis, home care, etc.). Use will be made of a satellite-based infrastructure interoperating with different broadband and narrowband wireless terrestrial systems which permit to offer first-aid services to mobile vehicles (ambulance, ship, airplanes).

TELESAL is a joint venture among Kell, Space Engineering and Telbios, in which eleven other organizations participate as subcontractors. Health care administrations participate in *TELESAL* as pilot users. The project, which began in June 2006 and will last 34 months, encompasses the definition of user requirements, the design of the network infrastructure, the hardware procurement and system integration and finally the experimental pilot service provision phase with the involvement of a significant number of public, private and transportable hospitals, research centres, specialistic centres, doctors and patients.

After the validation of these services performed by the involved operators, the participants will establish a consortium, possibly with the participation of institutional users that will run the services and their diffusion in Italy and abroad.

For the purposes of the pilot phase the following facilities will be set-up:

- a meshed sub-network among some of most important Italian Medical Centres (Hospital, Clinical Centre, Medical Research Centres, etc.);
- a collection of star sub-networks, each one tailored for a specific application; all the *TELESAL* sub-networks will have a gateway toward the meshed network used for connecting Medical Centres;
- a capillary distribution network for broadcasting piece of information to large population of operators (i.e., doctors) or users (i.e., patients);
- where convenient, some communication wireless terrestrial tails used to share among many fixed users a single satellite access point.

All system activities will be coordinated by three Service Centres, each one specialised for a particular class of application. The Service Centres will be complemented by:

- a common Call Centre intended to interface user communities in the most appropriate and efficient way;
- a Supervision Service Centre mainly dedicated to the management of network resources shared among different applications, and playing the role of central node of a web-based infrastructure intended to federate, in interoperable manner, some telemedicine systems already existing and operational in Italy, at local and regional level.

The applications (emergency, home care, screening, telemedicine aboard ships and airplanes) will be developed employing heterogeneous software technologies and programming languages, and using the IP network protocol as the unifying paradigm. Large use of relational databases will be done.

27.6 Conclusions

The technologies and applications described in this chapter are good examples of innovative approaches developed in civilian projects that can also be utilized in military contexts. They also demonstrate that the small-medium enterprises can provide a tangible contribution to progress in space communications.

Chapter 28
ELSAG DATAMAT Outlook on Dual Use

Claudio De Bellis

ELSAG DATAMAT

Elsag Datamat Spa is the ICT company of Finmeccanica, established on 1st August 2007 by merging Elsag Spa and Datamat Spa, both belonging to **Finmeccanica Group**, Italy's leading industrial group in the advanced technology sector.

Finmeccanica Spa is the industrial and strategic holding company which coordinates the work of group business, operating in aeronautics, helicopters, space, defence, energy and transport, with around 60,000 staff, 13 Billion Euro revenues and 14% R&D investments.

With almost 4,000 staff and over 650 Million Euro revenues, **Elsag Datamat** provides in-depth knowledge of **automation**, **security**, **defence** and **space**, **transport** and **information technology** processes, for advanced technology products and services.

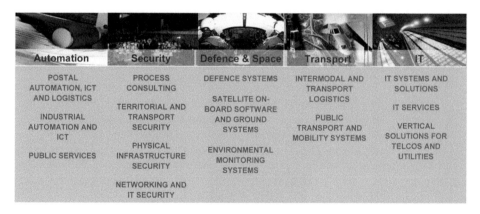

Automation	Security	Defence & Space	Transport	IT
POSTAL AUTOMATION, ICT AND LOGISTICS	PROCESS CONSULTING	DEFENCE SYSTEMS	INTERMODAL AND TRANSPORT LOGISTICS	IT SYSTEMS AND SOLUTIONS
INDUSTRIAL AUTOMATION AND ICT	TERRITORIAL AND TRANSPORT SECURITY	SATELLITE ON-BOARD SOFTWARE AND GROUND SYSTEMS	PUBLIC TRANSPORT AND MOBILITY SYSTEMS	IT SERVICES VERTICAL SOLUTIONS FOR TELCOS AND UTILITIES
PUBLIC SERVICES	PHYSICAL INFRASTRUCTURE SECURITY	ENVIRONMENTAL MONITORING SYSTEMS		
	NETWORKING AND IT SECURITY			

250 staff within Elsag Datamat are working for space, including its subsidiaries **Dataspazio Spa** and **Space Software Italia Spa (SSI)**, covering all SW-based systems and applications for:

Space Segment:

–   On-board **data handling SW** for Satellites (Sicral, Cosmo/SkyMed)
–   On-board **mission data base** for Space Station ISS (Columbus)

- On-board **mission SW** for Space Vehicles (Hermes, ATV, LYRA)
- On-board **flight SW** for Unmanned Vehicles (UAV)
- Payload and equipment **monitoring & control** (Cosmo/SkyMed, ATV)
- **SW quality assurance** and **configuration management** (VEGA small launcher)

Ground Segment:

- **Satellite Control Centres** (for all Italian satellites: from Sirio1 to Italsat and Cosmo/SkyMed, SAX, and some European: Artemis, XMM, Cryosat, Goce, . . .)
- **Launch Control Systems** (for VEGA small launcher)
- **Simulators** of Satellite, Payload, Ground Segment (SAX, Cosmo/SkyMed, GALILEO, . . .)
- **Advanced Networking** and **Monitoring & Control** (HiSCEK, HiSEEN)
- **Mission Exploitation Centres**
  - *Scientific* (SAX, XMM . . .)
  - *Telecommunications* (Italsat, Sicral, . . .)
  - *Positioning / Navigation* (EGNOS, GALILEO, . . .)
  - *Earth Observation* (Helios, Cosmo/SkyMed, Envisat, MUIS, current and new ESA Missions, GMES, . . .):
    - *Distributed User Services*
    - *Multi-Mission Infrastructures*
      - Mission and Payload Planning and Control
      - Data Ingestion, Processing and Products Generation
      - Data Search, Access, Archiving, Ordering and Distribution
      - Distributed Information Management

- *Ground Segment Engineering*
  - ■ G/S Design and Development
  - ■ G/S Assembly, Integration and Validation
  - ■ G/S Engineering Facilities and Services
  - ■ G/S Operations Engineering Support

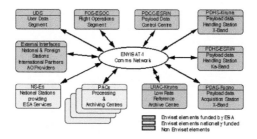

User Segment:

- **Operational Systems and Applications** for:
  - *Meteorology*:
    - o Weather Radar Network C&C and processing (DataMet Suite)
    - o Meteorological Satellite Exploitation (Data Ingestion and Processing Chain for Meteosat MOP, MTP, MSG, TYROS_N/NOAA)
    - o End-to-end data processing and forecast (National Meteo Service Management System)
    - o Rapid deployable network with TCP/IP protocol (HF Emergency Networks)
  - *Environment Monitoring*
    - o Integrated solution for Flood Forecast and Management (M3Flood)
      - ■ *Emergency Management*
    - o On-field operations support based on ICT fixed and mobile equipment (EGERIS)
      - ■ *Security and Law Enforcement*
    - o Information Systems (for Europol, for Schengen enforcement)

o   Integrated Ballistic Identification System for forensics (IBIS)
o   Infrastructures Protection and Supervision (Global Area View —
     GAV)

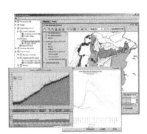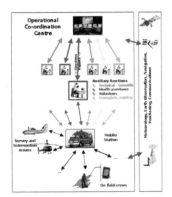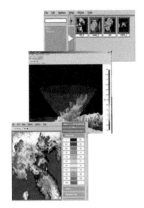

Pre-requisites for dual use are: coordination — collaboration — cooperation.

–   *Coordination* is an agreement to put own information at disposal of a partner
    for a specific scope, e.g. like:
    •   personal data: to enforce Schengen treaty, Europol, . . .
    •   communication links: Police both military and civilian, Coast Guard,
        Civil Protection, . . .
    •   meteorology: weather forecast, . . .
–   *Collaboration* is an agreement to share and assign duties and actions to reach
    a common goal;
–   *Cooperation* is an agreement to access the same data and operate a joint
    system, although for possible different aims and priorities.

Agreements for dual-use have been reached and even systems suited for dual use are
being implemented: then, what and where is the problem?

The major issue is ***EXPLOITATION***, i.e.: data access, archiving, processing,
dissemination.

For full exploitation, there is the need of a ***MULTI-MISSION INFRASTRUC-
TURE***, including the technologies to allow ***INTEROPERABILITY***, after having
esteblished well defined ***STANDARDS***.

Only if all this is achieved and implemented, then the exploitable ***DATA
FUSION*** will be possible.

Since at least 15 years, all major international Entities, like NASA, ESA, EC,
EUMETSAT, are investing significant efforts on technologies and infrastructures:

- Standard interoperable protocols for data access and exchange (Z39.50, CIP, GIP, GML, XML, SOA, . . .);
- Standard approach and methodology (INSPIRE, ORCHESTRA, . . .);
- GRID technology for shared archiving and processing;
- Ground Segment Multi-Mission Infrastructures.

The most popular dual-use application is surely *METEOROLOGY*, but more recently also *EMERGENCY MANAGEMENT, SECURITY and LAW ENFORCEMENT* imply and require the closest coordination among all involved Entities (e.g.: Civil Protection Authority, Coast Guard, Fire Brigades, Military Forces, . . .), integrating satellite and ground technologies, more specifically telecommunications, earth observation and positioning/navigation.

Elsag Datamat is one of the most active companies in Europe in all these fields, mastering state-of-the-art technologies and setting-up *Multi-Mission Management & Exploitation Centres* and *End-User Applications* specifically suited *for dual-use*, in summary:

- **Meteorology** (Satellite, Radar and Gorund Sensors): Data Processing Centres for Italian National Meteorological Service of Air Force, Regional Meteorological Services and Civil Protection Authority;
- **User Services**: Helios for MoD, Envisat for ESA;

- **Multi-Mission Facilities Infrastructures** (MMFI) for ESA: ODISSEO, FEOMI, GMES G/S, new ESA missions;
- **Interoperability**: CIP for EC, DAIL/HMA for ESA;
- **Secure communications**: military data link 11-16-22, multi-media communications VOIP on VHF.

Elsag Datamat is willing to put all its skills and background experiences at disposal of Customers, End-Users and large Prime Contractors.

Chapter 29
A Contribution from EADS Astrium

Michel FEUGA

EADS Astrium

29.1 Introduction

Space contribution to the European Security and Defence Policy (ESDP) is making today more and more sense and shall bring important added value for building up new enhanced capabilities needed for future civilian and military operations.

In this new context, where cost effective solutions have to be worked out, dual-use capabilities of both civilian and military existing and future programmes should be developed to a large extend, together with interoperability principles and associated standards.

While threats are evolving to become more diverse, less visible and less predictable, defence, or more generally "security" needs, become more and more demanding; access to as many space infrastructures as available should then ease emerging of a new set of applications thanks to enhanced overall performances in the fields of earth observation (enhanced resolution, reduced time to revisit), telecommunications (bandwidth availability) or navigation (accuracy, integrity).

Defence applications should take benefit of the combination of civilian space assets for general use with military ones for strategic areas (ex: large swath from commercial market together very high optical resolution from military domain for earth observation), while the development of new capabilities for civilian applications should be boosted with the progress achieved on military space infrastructures.

Since several years, EADS Astrium has been involved in the delivery of programmes or services in the field of earth observation and telecommunications with either dual use capability or built-in duality; three examples of such background are given hereafter, providing they will help in understanding benefits of dual use technologies and applications.

29.2 Dual Use for On Demand Space Telecommunication Capacity

EADS Astrium holds, since 2005, a framework agreement ASTEL-S Convention (Figure 29.1) with the French Ministry of Defence to provide the French armed forces with satellite telecommunications services in both civilian Ku (between

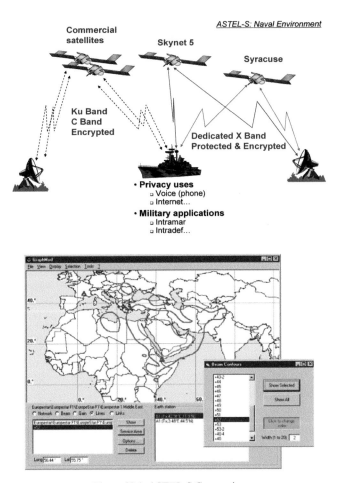

Figure 29.1 ASTEL-S Convention

Europe and external theatres of operations), and C bands and military band (SHF), for a renewable period of two years.

The applications focus on land-based, naval and airborne activities and the coverage of world-wide services (video-conferencing, Internet, data exchange).

When capacity is needed to complement the French Syracuse military satellite system, dedicated satellite communications capacity is provided by EADS Astrium and its ASTEL-S partners through a 'one-stop-shop' approach which enables an optimally tailored response to meet requirements as they occur.

EADS Astrium, prime contractor, has teamed with satellite capacity broker and supplier London Satellite Exchange (LSE) Ltd, whose operational base, despite the company's name, is in Toulouse (France).

A key factor in the EADS Astrium/LSE (Figure 29.2) offer to the MoD is the central function of LSE's highly elaborated database which contains details of more than 180 commercial telecommunications satellites (and their 5,500 transponders),

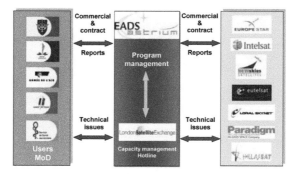

Figure 29.2 EADS Astrium/LSE

with specifications of their operational capacity, frequencies and range covered and availability at any given time, a unique tool for an instant overview of what resources can be called upon to make a perfect, cost-efficient match with requirements.

Thanks to a single unified interface for all contractual, commercial and technical issues, service delivery includes engineering, regulatory support functions, booking office and customer care (24x7 hotline).

For ASTEL-S purposes, EADS Astrium has concluded agreements with almost 10 major satellite operators for C- and Ku-band, and with its subsidiary, Paradigm Secure Communications, for X-band.

This perfect example of dual use service allows efficient and optimized access to a wide range of available space telecommunication capacity for prompt on-demand reactivity.

29.3 SPOT5 HRS, A First Step Towards Dual Programmes

HRS (High Resolution Stereoscopic - panchromatic mode), carried on SPOT5[1], is the first high-resolution sensor on the SPOT constellation that allows the acquisition of stereo images in along-track direction, using two telescopes pointing about 20 degrees forward and backward. In a single pass, the forward-pointing camera acquires images of the ground, and then the rearward-pointing camera covers the same strip 90 seconds later. HRS is thus able to acquire stereo images almost simultaneously to map relief, produce Digital Elevation Models (DEMs) of wide areas and generate high quality ortho-rectified products.

Given that military operations rely more and more upon accurate, complete and up to date geographical data and as there was a need, in the private sector, for a geometrical reference when no reliable map is available to set up automatic

---

[1]SPOT5 is operational since July 2002; it belongs to the SPOT "optical" constellation and was developed under EADS Astrium prime contractor ship for CNES (Centre National D'Etudes Spatiales). SPOT5 is flying along a near-polar, near-circular and Sun-synchronous orbit at a mean altitude of 832 km, an inclination of 98.7 degrees and a mean revolution period equal to 101.4 minutes. The SPOT satellites orbit the same ground track every 26 days.

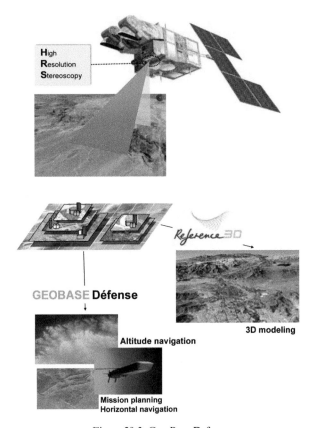

Figure 29.3 Geo Base Defene

production of DEMs and ortho-images; it was decided, in the late nineties, to launch HRS instrument as a full dual programme, 54% funded by the private sector (CNES, EADS Astrium, Spot Image), 46% by the French MoD.

Thanks to HRS acquired images and through a partially common production chain, two final data bases are produced: "Reference3D" for the private sector and "Geobase Défense" for the defence world.

Reference3D package contains an accurate DEM, HRS ortho-images and some useful metadata describing the DEM quality. Produced by Spot Image in partnership with the French Mapping Agency (IGN), Reference3D is available virtually everywhere and offers a good accuracy with no need for ground control points at all.

Geo Base Defene (Figure 29.3) is the primary base layer generated under the DNG3D programme; it covers some 30,000,000 Km2 with digital terrain models to the DTED2 standard, an ortho-image carpet (at a pitch of 5m and using ARC projection) enabling the location of any type of image source (by automatic resetting), and XML metadata to the ISO 19115 standard. Resulting location accuracy is compatible with STANAG 4294 (relative to GPS).

"Reference3D" and "GeoBase Défense" are twin products with some differences: same data terrain elevation data level 2 layer, same accuracy, same quality, same production process (Spot Image & IGN) but dedicated ortho-image projections, specific framing principles and different combination and presentation of the quality layers (metadata).

This second example of dual use programme has shown that significant savings came from the duality of the programme itself (single development) but also from joint exploitation for part of the production chain during its operational life. On top of that, thinking dual from the very beginning for this programme has for sure been key to make it happen.

29.4 PLEIADES Dual System Concept

Figure 29.4 is a CNES <u>dual</u> programme designed as the follow-on to the SPOT family of low Earth orbit (LEO) observation satellites. The high-resolution Pleïades global constellation (a pair of satellites for sub-metric-resolution observation) is based on smaller, cheaper and more agile satellites, and was initially part of the ORFEO programme as the optical component, the Italian Cosmo-Skymed system being the radar one.

Later on, CNES signed also Pleiades cooperation agreements with SNSB (Swedish National Space Board) with INTA (Instituto Nacional de Técnica Aeroespacial) of Spain, with ASA (Austrian Space Agency) of Austria, and with BELSPO (Belgian Science Policy Office) of Belgium. Pleiades becomes, within this context, a multi-mission concept and a partnership dual programme.

Pléiades first launch is planned for end 2009 with the second one year later. EADS Astrium is prime contractor for the Pleïades satellite bus, including all functions dedicated to satellite control and monitoring as well as payload data handling and transmission, and is responsible for software development and satellite validation. After being involved during ground segment engineering and development phases, EADS Astrium has been recently chosen as prime for ground system integration / validation and deployment activity.

Figure 29.4 Pleiades

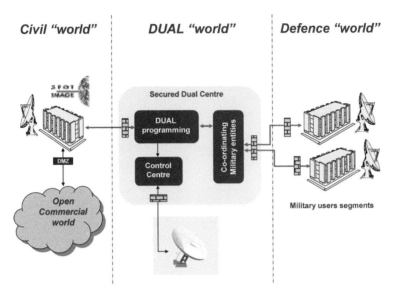

Figure 29.5 Dual Applications

Pleiades dual system was designed considering requirements coming from both Defence and Civil worlds, most of them being common (Figure 29.5):

- Among common requirements most significant are large quantity of images to be acquired, flexibility (sizes, resolutions, agility), reactivity (response / revisit times), availability and sustainability over 10 years.
- Defence specific requirements are mainly related to ability to handle high priority access (responsive tasking), classified tasking, classified products and secured communications
- Civil related requirements are aiming at offering access to a wide range of users community such as institutional (GMES), scientists or commercial thanks to open access via public network (Internet)

The Pleiades ground segment is composed of a set of user centres located in France, Sweden, Spain and (optionally) Italy, all of them being interfaced with a dual centre in France. A user centre may be for civil or defence purposes. Duality implies some defence specific requirements, in particular security related ones. Secured communications between civil and defence worlds are implemented thanks to a dedicated equipment family; this set of equipments, developed by EADS Astrium, acts as an extremely robust firewall and is used to segregate civil open world from defence one; capability which is key when we talk about dual use.

29.5 Dual Use Systems: Guidelines for the Future

The examples quickly described here above show how EADS Astrium has been involved since several years in the development of dual use programmes, either at space segment level or for ground infrastructures. From this strong background we have in particular learned that, as soon as dual use is anticipated or even envisaged, a lot of issues have to be solved in the frame of ground system design activity. Combining military and civilian uses for a single (or a set of) sensor(s) always results in specific requirements for managing sensor sharing and security issues; in that frame, EADS Astrium has developed and is continuously enhancing expertise in key areas such as multi-mission planning (direct tasking, priorities policies, responsive mode...) and "safe" resources sharing (privacy and data segregation, security and anti-intrusion). On top of that, future programs such as MUSIS, mixing multi nation co-operation, multi sensors sharing and dual use, will need strong expertise and background with respect to "secure" interoperability issues; past and current strong involvement in several studies and developments dealing with mission and data exploitation interoperability and security handling (HMA, Comu...) positions EADS Astrium as a key player ready to propose innovative and efficient solutions for ground systems design, development and deployment.

Chapter 30
Cell Broadband Technology: An Overview

Elisa Tonello

IBM Italia S.p.A. — Italy

30.1 Cell Broadband Engine Technology

The Cell Broadband Engine technology was conceived through the research efforts by Sony Computer Entertainment Inc., Toshiba and IBM. The STI alliance started in 2000 and derived the first implementation of the processor, finding its first installation in Sony PlayStation3, in 2005. The architectural concept had to satisfy challenging objectives in terms of performance, real time responsiveness and power consumption, giving rise to a flexible, powerful solution which could target the multimedia application requirements and a broad range of computing needs beyond them.

30.1 Power Processing Element

30.2 Architecture Basics

The Cell B.E. design combines an IBM 64-bit PowerPC based processor (Power Processing Element - PPE (Figure 30.1)) running at 3.2 GHz with eight 128-bit SIMD (Single Instruction Multiple Data) cores, the Synergistic Processing Elements (SPEs), all connected via a high-speed bus. The PPE is responsible for running the operating system and distributing tasks among the eight SPEs, which take in charge the compute intensive parts of the OS and the applications. This architectural choice, giving full 32/64-bit Power Architecture compatibility, provides a powerful base for porting and multi-parallel application design on Cell. With its RISC, in-order, dual-threaded architecture, the PPE is a low consuming processor that can fetch four instructions and issue two at once. It includes two 32 KB L1 caches and one 512 KB L2 8-way write-back cache, and it can theoretically deliver double precision 6.4 Gflops and 25.6 Gflops in single precision.

The eight SPEs are in-order vector processors, with 128x128 bit registers, capable of performing SIMD instructions on 4 single precision values. At 3.2 GHz, each SPE can deliver up to 25.6 Gflops in single-precision, performing a maximum of 64 single-precision floating point operations per clock cycle, with the majority of the computational power resting on the SIMD units.

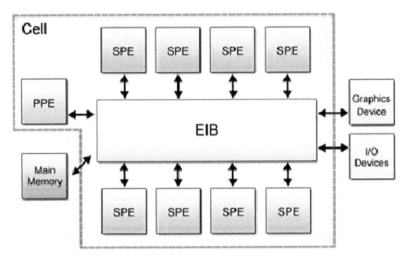

30.2 Element Interconnect Bus (EIB)

The SPE implements a new instruction-set architecture optimized for performance on computing-intensive applications. They operate on a local store memory (256 KB) that stores instructions and data, which are transferred between this local memory and system memory by asynchronous coherent DMA commands, executed by the memory flow control unit included in each SPE. Each SPE supports up to 16 DMA commands. As these coherent DMA commands use the same translation and protection governed by the page and segment tables of the Power Architecture as the PPE, addresses can be passed between the PPE and SPEs, and the operating system can share memory and manage all of the processing resources in the system in a consistent manner.

The PPE and SPEs are connected by a unit called the Element Interconnect Bus (EIB) (Figure 30.2), providing a single address bus and four 16-byte wide data rings. Data can be transmitted clockwise on two of the rings and counterclockwise on the other two. As each unit on the EIB can simultaneously send and receive 16 bytes of data every bus cycle, this gives a theoretical peak data bandwidth for the EIB as 96 bytes per processor clock cycle.

An integrated memory controller provides a high-performance, low-latency controller interface to XDR-based DRAM memory systems. External memory is accessed via two XIO channels operating at 3.2 GHz. Each XIO channel has separate, independent read and write request queues with an arbiter that alternates between them.

Moreover, Cell is designed with a high-bandwidth configurable I/O interface, a Rambus FlexIO controller providing seven transmit and five receive links. Each link is 1 byte wide. This translates to a peak bandwidth of 25 GB/s inbound and 35 GB/s outbound. These links can be configured as two logical interfaces.

All of these features translate into a multi-core processor capable of delivering extremely high performance in a wide range of application areas.

30.3 Cell Programming

Cell was designed to achieve outstanding results in performance while maintaining programmability. A full-level exploiting programming for Cell should play with multithreading, multiple core parallelization, cache management, SIMD instructions and DMA management, to which we can add high level parallelization in a multiple Cell system. Cell processor software development in the C/C++ language is supported, with a rich set of language extensions for SIMD operations and intrinsics to assembly instructions. By building on Power technology, applications can run without modification, and this allows for a staged approach, where code is developed and then SIMD-vectorized, before performance is enhanced by using the synergistic processors.

In order to extend the reach of this architecture, and to foster a software development community who could work on optimizing applications for this architecture, an open, Linux-based, software development environment was developed.

Cell scalability and multilevel parallelism make it an effective platform for a wide range of situations, starting from digital and multimedia applications but going far beyond. Ray-casting, online gaming with real time physics simulation, cloth simulation are only the first examples in which Cell power has been exploited, demonstrating the platform remarkable possibilities. Interest was generated in several markets, and investigations have been made for applications in financial analytics, medical digital imaging, ray tracing, and so on.

30.4 Cell Systems

IBM BladeCenter QS20, the first Cell-based blade, equipped with two Cell processors and 1GB RAM memory, is available since July 2006. Its successor, BladeCenter QS21, has been just announced, with duplicate RAM and the same performance, 460 Gflops single precision and 42 Gflops double precision, in half the space. They can be integrated in cluster configurations and are included in IBM Cluster 1350 portfolio.

Along with the hardware offering, the IBM Software Development Kit for Multicore Acceleration has been released, containing several instruments to develop applications for Cell hardware on a variety of platforms, including x86, x86_64, PowerPC, and Cell Blades, and equipped with rich documentation and tutorials. The development kit contains compilers for both PPE and SPE, software libraries as SIMD mathematical libraries, Basic Linear algebra Subroutines, and a framework for data and task parallel application and library programming. To increase ease of programming and developers productivity, the SDK includes sample sources, performance tools and Eclipse-based integrated development environment (IDE) plug-ins for building, compiling and debugging applications leveraging the compilers. A full-system simulator is also available for Cell development investigation and test.

Some open source components of the SDK are distributed by the Barcelona Supercomputing Center, whose commitment in the Cell project is also addressed to the development of a framework for the automatic exploitation of parallelization.

Cell is also a scalable system architecture. IBM, together with Los Alamos National Labs, will be realizing a hybrid, Cell-based supercomputer, named Road-Runner. It will involve over 16000 IBM System x servers based on AMD Opteron technology, used in conjunction with over 16000 Cell Blades, in a high-bandwidth, low latency connection, for a new heterogeneous design in supercomputing. It is expected to cross the petaflop barrier.

30.5 Conclusions

The Cell technology, with a general-purpose system flexibility and the power of an accelerator processor, the growing library and tool environment and an adequate skill development, can rise interest and propose its powerful parallelism paradigm in a wide range of markets.

30.6 Resources

IBM Official site on Cell BE http://www-03.ibm.com/technology/cell/index.html offers a suitable starting point.

Cell BE-related articles, discussion forums, downloads, and more could be found at the IBM developerWorks site http://www.ibm.com/developer/power/cell.

The Cell Broadband Engine Documentation http://www-01.ibm.com/chips/techlib/techlib.nsf/products/Cell_Broadband_Engine lists specifications, user manuals, and more.

Chapter 31
A Contribution from OSCAR

Domenico D'Angelo

Oscar

31.1 The Company

Set up in June 1999, Oscar specialises in ICT services. The company is founded upon business and social values held in common by the original shareholders, and laid out in the document called "Charter of Corporate Values". Our common belief was, and remains to this day, that the respect and enhancement of the personal and professional skills of our workforce, united to the strict observance of commercial correctness, are far from inimical to profit-making.

Proof positive of which is the fact that in these first few years of existence Oscar has expanded its activities, carried out projects of ever-increasing level, acquired prestigious clients, and strengthened its capital structure.

31.2 Security and Data Protection

The fast-developing data security sector is major strategic priority of Oscar. Our earliest experiences with our customers have made us realize that while total and absolute security is probably unattainable, there are viable, cost-effective and above all ever-evolving solutions for protecting systems databases.

Today Oscar takes pride in the substantial national and international endorsements of the company and of its employees, whose proven professional expertise remains the bedrock of its activity in the area of data security. Oscar focuses on both defence and civil market.

31.3 The Offer

Oscar's corporate strategy has given priority the developing diverse types of activities while maintaining high standards, as confirmed by the consistently positive feedback of its business clients. During the initial contact phase with customers and institutions, paramount attention is given to project arising from innovative ideas, whether put forward by Oscar of its prospective customer. The subsequent implementation of these ideas is carried out in close partnership between the two.

The company's main areas of operations are the following:

- Software products design, implementation and distribution.
- Fully integrated solutions projects (hardware and software).
- Advisory and support services to corporate and public entities.
- ICT Security.

31.4 Our Security Focus

The subject of the conference is the Dual Use (military and civilian) of aerospace technologies. Even if what we're going to present isn't exclusively an aerospace technology , however is also necessary in the aerospace sector. On terms of Dual Use we like to underline that the presented solution is the product of the cooperation between two European companies, the Italian OSCAR and the Dutch OSPL. In particular we want to talk about TEMPEST technology. TEMPEST is the code name of the study of electromagnetic emissions of all electronic apparatus processing data. From scientific studies it was proved that all electrical equipment produces unintentional readable electromagnetic emissions, and it's possible to capture these emissions and decode information. On terms of collaboration between the two companies, OSCAR's role is to research for customer's needs and to offer new solution that will be analysed, tested and certified from OSPL.

31.5 About OSPL

OSPL Netherlands is a manufacturer and supplier of TEMPEST, Rugged and secure equipment to the Ministry of Defence and Government Agencies both nationally and internationally. OSPL is in business since over 25 years. As part of its core capability, OSPL operates and manages a TEMPEST test chamber which is approved and supported by the Netherlands National Comsec Agency, [NL NSA].

Utilising the latest FSET receiver technology, together with a current facilities qualification certificate this chamber is capable of performing TEMPEST testing to the highest NATO TEMPEST standard, SDIP-27 level A. OSPL is an ISO 9001 registered company and as such provides the highest standards of quality assurance and project control. Until today the sensibility to TEMPEST problematic was exclusively felt by the military sector, intelligence and from civil companies that develop solutions for the defence sector. In the last year the necessity to protect sensible information (considered sensible from the company) has defined the extension of TEMPEST solution also on other productive and financial sectors.

> Set Up in 1999

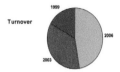

> Specialized in IT Security Sector since 2003
  •INFOSEC / COMSEC
  •Security Assessment & Evaluation / Risk Analysis
> International Strategic Alliances

  •Checkpoint
  •OSPL → TEMPEST

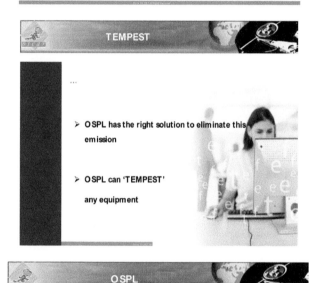

> OSPL has the right solution to eliminate this emission

> OSPL can 'TEMPEST'

any equipment

OSPL

What can OSCAR & OSPL do for you?

> **Government & Civilian**

> **Anti-terrorism**

> **Defence**

  » NATO qualifications

  » Accreditations

31.6 Conclusion

The Oscar's main objective can be summarized in the following statements:

1. **Enter the Italian Civilian Market**, because the data protection needed in the civilian industry is certainly not less than the military one.
2. **Enter the Italian Military Market in full compliance with a "ANS" Security qualification process:** the Italian security agency gives specific and more restricted procedures for the use of TEMPEST equipment.
3. **Have a fair competition with existing players:** at the present time in Italy exist national manufactures and distributors of TEMPEST solution at the best level. A fair competition, both for technological and economical issues, is the positive perspective for the market and the end users.

Chapter 32
A Contribution from Lockheed Martin

Mesut Ciceker

Lockheed Martin Corporation

Lockheed Martin Space Company is one of the four major lines of businesses of Lockheed Martin Corporation. Space Company has wide range of designs and products to serve for both civil and military purposes in spacecraft and space transportation.

32.1 Spacecraft - The Versatile A2100 Spacecraft Family

The A2100 design (Figure 32.1) is highly modular at the subsystem and component levels. It features a major reduction in parts - simplifying construction, increasing on-orbit reliability and reducing weight and cost. Lightweight all-composite materials increase strength, minimize thermal distorsions and reduce launch costs.

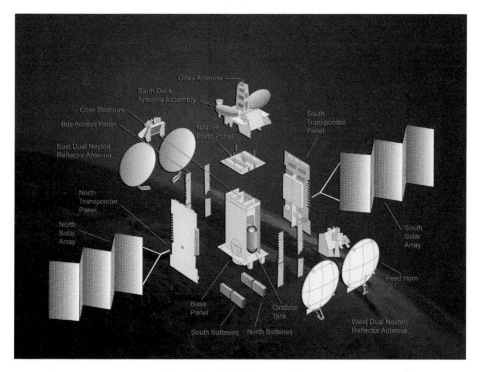

Figure 32.1 A2100 Expanded View

A2100 is the culmination of 50 years of Lockheed Martin investment in innovation. Since SATCOM 1, Lockheed Martin's first three-axis -stabilized geosynchronous satellite, Lockheed Martin has built and launched more than 90 communication satellites.

Lockheed Martin continues to invest in enabling technologies such as high-density electronics to increase power capability, advanced antennas to increase payload communications capacity and more efficient propulsion systems for increased orbital efficiency and reduced launch cost. Systems currently in production or under development will use A2100 satellites to provide advanced services such as video on demand, video teleconferencing, telemedicine, high-speed internet access and multimedia services.

The Lockheed Martin A2100 geosynchronous spacecraft series is designed to meet a wide variety of telecommunications needs including Ka-band broadband and broadcast services, fixed satellite services in C-band and Ku-band, high-power direct broadcast services using the Ku-band frequency spectrum and mobile satellite services using UHF, L-band, and S-band payloads. The A2100's modular design features a reduction in parts, simplified construction, increased on-orbit reliability and reduced weight and cost. A2100 spacecraft is widely accepted and being used by the major commercial satellite operators worldwide.

Figure 32.2 SES Americom AMC 16   Figure 32.3 SES Sirius

The A2100 spacecraft's design accommodates a large range of communication payloads. This design modularity also enables the A2100 spacecraft to be configured for missions other than communication. The A2100 design is currently being adapted for geostationary earth orbit (GEO)-based earth observing missions and is currently the baselined platform for Lockheed Martin's Geostationary Operational Environmental Satellite Series-R (GOES-R) proposal. The A2100 also serves as the platform for critical government communications programs including Advanced Extremely High Frequency (AEHF) (Figure 32.5) and Mobile User Objective System (MUOS) (Figure 32.4) and is the foundation for Lockheed Martin's Transformational Satellite Communications System (TSAT) offering.

Figure 32.4 MUOS

Figure 32.5 AEHF

32.2 ATLAS Launch Vehicle

The Atlas/Centaur launch vehicle is manufactured and operated by Lockheed Martin to meet commercial and government medium, intermediate, and heavy space lift requirements. The Atlas program began in the mid-1940s with studies exploring the feasibility of long-range ballistic missiles. The Atlas launch vehicle family has evolved through various United States Air Force (USAF), National Aeronautics and Space Administration (NASA), and commercial programs from the first Research and Development (R&D) launch in 1957 to the current Atlas V configurations. More than 565 Atlas vehicles have flown to date.

Versions of Atlas Boosters were built specifically for manned and unmanned space missions, including the pioneering Project Mercury manned launches that paved the way toward the Apollo lunar program. The addition of the high-energy Centaur upper stage in the early 1960s made lunar and planetary missions possible. In 1981, the Atlas G booster improved Atlas/Centaur performance by increasing propellant capacity and upgrading engine thrust. This baseline was developed into the successful Atlas I, II, Atlas IIA, IIAS, IIIA and IIIB launch vehicles.

Atlas V continues the evolution of the Atlas launch vehicle family. Today, as the world's most successful launch vehicle, the Atlas V is offered in a comprehensive family of configurations that efficiently meet spacecraft mission requirements (Figure 32.6 and 32.7).

The Atlas system is capable of delivering a diverse array of spacecraft, including projected government missions to Low Earth Orbit (LEO), heavy lift Geosynchronous Orbits (GSO), and numerous Geostationary Transfer Orbits (GTO). To perform this variety of missions, Lockheed Martin combines a Common Core

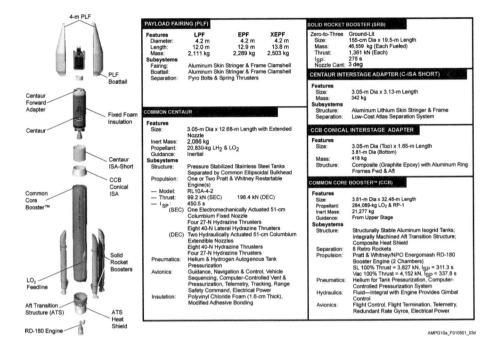

Figure 32.6 Atlas V - 4 m fairing

Booster (CCB) powered by a single RD-180 engine with a standard Atlas V 4m Long Payload Fairing (LPF), Extended Payload Fairing (EPF) or Extended EPF (XEPF) to create the Atlas V 400 series. For larger and heavier spacecraft, the Atlas V 500 series combines the CCB with a 5m Long Payload Fairing. The Atlas 400 and 500 series include a Common Centaur that can be configured with either a single or dual engine, depending on mission requirements.

The Atlas V system (Figure 32.8 and 32.9) provides increased reliability over its predecessors. The increased reliability is achieved through a simplified design that incorporates fault avoidance, fault tolerance, and reduction of Single Point Failures (SPF). The robustness of the Atlas V system is enhanced by the use of common system elements assembled into a family of vehicles that satisfy a wide range of mission requirements while providing substantial performance margins. In addition to common elements, the Atlas V system features improved structural capability allowing it to withstand worst-case day-of-launch winds. The result is increased launch availability.

There were eleven Atlas V missions to date (since August 2002), and more than 50% of those flights were for commercial satellite operators.

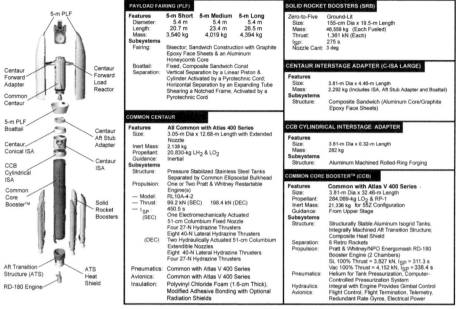

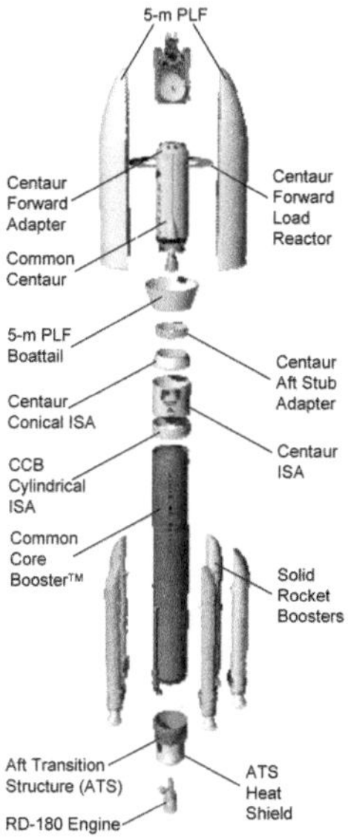
5-m PLF
Centaur Forward Adapter
Centaur Forward Load Reactor
Common Centaur
5-m PLF Boattail
Centaur Aft Stub Adapter
Centaur Conical ISA
Centaur ISA
CCB Cylindrical ISA
Common Core Booster™
Solid Rocket Boosters
Aft Transition Structure (ATS)
ATS Heat Shield
RD-180 Engine

PAYLOAD FAIRING (PLF)

Features	5-m Short	5-m Medium	5-m Long
Diameter:	5.4 m	5.4 m	5.4 m
Length:	20.7 m	23.4 m	26.5 m
Mass:	3,540 kg	4,019 kg	4,394 kg
Subsystems			
Fairing:	Bisector; Sandwich Construction with Graphite Epoxy Face Sheets & an Aluminum Honeycomb Core		
Boattail:	Fixed, Composite Sandwich Const		
Separation:	Vertical Separation by a Linear Piston & Cylinder Activated by a Pyrotechnic Cord; Horizontal Separation by an Expanding Tube Shearing a Notched Frame, Activated by a Pyrotechnic Cord		

COMMON CENTAUR

Features	**All Common with Atlas 400 Series**
Size:	3.05-m Dia x 12.68-m Length with Extended Nozzle
Inert Mass:	2,138 kg
Propellant:	20,830-kg LH_2 & LO_2
Guidance:	Inertial
Subsystems	
Structure:	Pressure Stabilized Stainless Steel Tanks Separated by Common Ellipsoidal Bulkhead
Propulsion:	One or Two Pratt & Whitney Restartable Engine(s)
— Model:	RL10A-4-2
— Thrust:	99.2 kN (SEC) 198.4 kN (DEC)
— I_{SP}:	450.5 s
(SEC)	One Electromechanically Actuated 51-cm Columbium Fixed Nozzle Four 27-N Hydrazine Thrusters Eight 40-N Lateral Hydrazine Thrusters
(DEC)	Two Hydraulically Actuated 51-cm Columbium Extendible Nozzles Eight 40-N Lateral Hydrazine Thrusters Four 27-N Hydrazine Thrusters
Pneumatics:	Common with Atlas V 400 Series
Avionics:	Common with Atlas V 400 Series
Insulation:	Polyvinyl Chloride Foam (1.6-cm Thick), Modified Adhesive Bonding with Optional Radiation Shields

SOLID ROCKET BOOSTERS (SRB)

Zero-to-Five	Ground-Lit
Size:	155-cm Dia x 19.5-m Length
Mass:	46,559 kg (Each Fueled)
Thrust:	1,361 kN (Each)
I_{SP}:	275 s
Nozzle Cant:	3 deg

CENTAUR INTERSTAGE ADAPTER (C-ISA LARGE)

Features	
Size:	3.81-m Dia x 4.46-m Length
Mass:	2,292 kg (Includes ISA, Aft Stub Adapter and Boattail)
Subsystems	
Structure:	Composite Sandwich (Aluminum Core/Graphite Epoxy Face Sheets)

CCB CYLINDRICAL INTERSTAGE ADAPTER

Features	
Size:	3.81-m Dia x 0.32-m Length
Mass:	282 kg
Subsystems	
Structure:	Aluminum Machined Rolled-Ring Forging

COMMON CORE BOOSTER™ (CCB)

Features	**Common with Atlas V 400 Series**
Size:	3.81-m Dia x 32.46-m Length
Propellant:	284,089-kg LO_2 & RP-1
Inert Mass:	21,336 kg for 55Z Configuration
Guidance:	From Upper Stage
Subsystems	
Structure:	Structurally Stable Aluminum Isogrid Tanks; Integrally Machined Aft Transition Structure; Composite Heat Shield
Separation:	8 Retro Rockets
Propulsion:	Pratt & Whitney/NPO Energomash RD-180 Booster Engine (2 Chambers) SL 100% Thrust = 3,827 kN, I_{SP} = 311.3 s Vac 100% Thrust = 4,152 kN, I_{SP} = 338.4 s
Pneumatics:	Helium for Tank Pressurization, Computer-Controlled Pressurization System
Hydraulics:	Integral with Engine Provides Gimbal Control
Avionics:	Flight Control, Flight Termination, Telemetry, Redundant Rate Gyros, Electrical Power

AMPG10a_F010501_04e

Figure 32.7 Atlas V - 5 m fairing

Figure 32.8 Atlas V RD 180 Engine

Figure 32.9 Atlas V (5 m Fairing) Launch

Index List